FUELS OF OPPORTUNITY:
CHARACTERISTICS AND USES IN COMBUSTION SYSTEMS

Elsevier Internet Homepage:

http://www.elsevier.com

Consult the Elsevier homepage for full catalogue information on all books, journals and electronic products and services.

Elsevier Titles of Related Interest

EL-MAHALLAWY AND HABIK
Fundamentals and Technology of Combustion
ISBN: 008-044106-8

Related Journals/Products

Free specimen copy gladly sent on request. Elsevier Ltd, The Boulevard, Langford Lane, Kidlington, Oxford, OX5 1GB, UK

Applied Energy
Biomass and Bioenergy
Combustion and Flame
Electric Power Systems Research
Energy
Fuel
Fuel and Energy Abstracts
Fuel Processing Technology
International Journal of Electrical Power and Energy Systems
Progress in Energy and Combustion Science
Renewable Energy

To Contact the Publisher

Elsevier welcomes enquiries concerning publishing proposals: books, journal special issues, conference proceedings, etc. All formats and media can be considered. Should you have a publishing proposal you wish to discuss, please contact, without obligation, the publisher responsible for Elsevier's Energy programme:

Henri van Dorssen
Publisher
Elsevier Ltd
The Boulevard, Langford Lane
Kidlington, Oxford
OX5 1GB, UK

Phone: +44 1865 843682
Fax: +44 1865 843931
E-mail: h.dorssen@elsevier.com

General enquiries, including placing orders, should be directed to Elsevier's Regional Sales Offices – please access the Elsevier homepage for full contact details (homepage details at the top of this page).

FUELS OF OPPORTUNITY: CHARACTERISTICS AND USES IN COMBUSTION SYSTEMS

David A. Tillman
Easton, Pennsylvania, USA

and

N. Stanley Harding
Salt Lake City, Utah, USA

2004
ELSEVIER
Amsterdam - Boston - Heidelberg - London - New York - Oxford
Paris - San Diego - San Francisco - Singapore - Sydney - Tokyo

ELSEVIER B. V.
Sara Burgerhartstraat 25
P.O. Box 211,
1000 AE Amsterdam
The Netherlands

ELSEVIER Inc.
525 B Street, Suite 1900
San Diego, CA 92101-4495
USA

ELSEVIER Ltd
The Boulevard, Langford Lane
Kidlington, Oxford OX5 1GB
UK

ELSEVIER Ltd
84 Theobalds Road
London WC1X 8RR
UK

First edition 2004

Library of Congress Cataloging in Publication Data
A catalog record is available from the Library of Congress.

British Library Cataloguing in Publication Data
A catalogue record is available from the British Library.

ISBN: 0-08-044162-9

The paper used in this publication meets the requirements of ANSI/NISO Z39.48-1992 (Permanence of Paper).
Printed in The Netherlands.

for Millie and Peggy

CONTENTS

PREFACE

Electricity deregulation, the new forces of the global economy, increased attention to environmental matters including but not limited to global warming, and numerous other regulatory pressures have increased the focus on delivering energy at the lowest economic and environmental cost. Such pressures have increased the focus on opportunity fuels. Opportunity fuels are those sources of energy that may result from other production processes, or may result from waste disposal practices, and may or may not enter the mainstream of energy commerce. For the most part they are sold and purchased outside the mainstream of energy commerce, although exceptions to this conclusion such as petroleum coke exist.

The array of opportunity fuels available is as broad and diverse as the ingenuity of the engineering community. It includes (not exhaustive) fossil-based fuels—petroleum coke, coal preparation wastes, coal-water slurries, methane from coal mining, and off-gas from petroleum refining—along with such biomass energy sources as sawdust and wood waste, herbaceous crops grown for energy purposes, methane-rich gases from the digestion of manure, and landfill gas. Within each of these broad categories are a plethora of specific fuels used by utilities and process industries.

Those of us who work in the area of fuels and combustion continuously face various issues associated with permitting, and using, the array of opportunity fuels produced and marketed to large energy consumers. Consequently we believed that there would be a use for a book dealing explicitly with these fuels. Our approach is to focus upon the large energy users—electricity generating stations and process industries—where the bulk of these opportunity fuels are used. We recognize that each and every specific opportunity fuel can not be dealt with explicitly. As a consequence we have chosen to address the more popular opportunity fuels, and to deal with examples that can be used to represent classes of opportunity fuels.

In dealing with the subject of opportunity fuels we recognize that one or more of these may become a mainstream fuel; Powder River Basin (PRB) subbituminous coal went from an opportunity fuel 40 years ago to a dominant energy source. PRB coal brought to the fore other subbituminous coals now available on the world market including the very low sulfur Indonesian subbituminous coals. We also recognize that some opportunity fuels discussed may fade back into obscurity as world economic and environmental conditions and regulations change.

Many individuals helped create this book. Those individuals include, in no particular order, Bruce Miller, Dr. Sharon Falcone Miller, and David Johnson of Pennsylvania State University; Martha "Bunni" Rollins, Lester Reardon, and Rick Carson of Tennessee Valley Authority; Dr. Sean Plasynski of USDOE-NETL; Evan Hughes and Dave O'Connor of EPRI; Glenn Holt, Tim Banfield, William Guyker (ret), Tom Nutter, Kathy Payette, and Erik Johnsson of Allegheny Energy; Kirby Letheby of Alliant Energy; Jerry Schmitt and Patricia Hus Peterson of NIPSCO; Joe Battista of Cofiring Alternatives; Don Kawecki and Byron Roth of Foster Wheeler; and Stephen Hoyle. Special thanks go to Dr. Richard Conn, Dr. Stefan Laux, and David Prinzing who provided extremely valuable review comments on all chapters in the writing process. Special thanks also go to our excellent editor, Ms. Victoria Thame. Without these individuals, we would have been unable to complete this project.

David A. Tillman

and

N. Stanley Harding

CHAPTER 1: OVERVIEW OF OPPORTUNITY FUELS

1.1. Introduction

Energy consumption in the world today is approximately 375 Exajoules or EJ (~375×10^{15} Btu), with the most significant energy-consuming regions including North America, Western Europe, and industrialized nations in Asia including Japan and South Korea [1]. North America also is the largest energy-producing region of the world, with other major energy producing regions including Asia and Oceania, Eastern Europe and the former Soviet Union, and the Middle East. Significantly, the regions of the world where energy consumption is highest—the industrialized regions—are those regions where there is a deficit between energy production and consumption. North America, for example, produces about 100 EJ, or quads, of energy annually; however it consumes about 115 EJ annually. Western Europe consumes about 70 EJ annually and produces about 45 EJ. Asia and Oceania produce about 75 EJ annually, and consume about 100 EJ/yr [1].

Energy consumption continues to grow throughout the world, with most of the growth occurring by use of petroleum and natural gas as is shown in Figure 1. Consequently, while there are regional balances that are seemingly close—North America, for example—there are substantial imbalances for such regions with respect to premium fuels. This exerts significant price pressures upon the energy market.

Energy consumption is unevenly divided across various sectors of industrialized economies, as it is unevenly divided across geographic regions. Electricity generation provides a critical example because of the significance of electrification in the economies of the USA and the

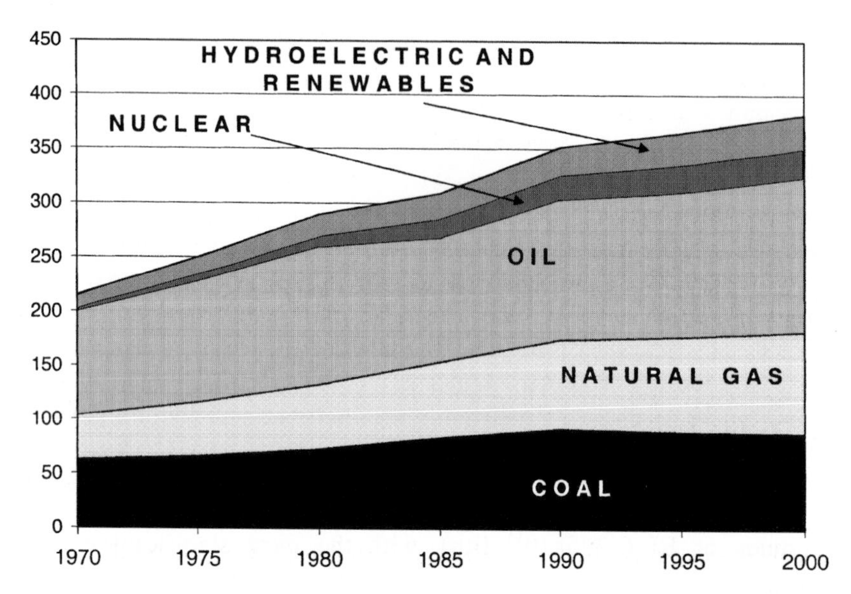

Figure 1. Historical Worldwide Energy Consumption by Fuel Type
Source [1]

industrialized nations. Electricity generation is dominated by the combustion of coal—particularly in the USA. Today some 56 percent of the electricity generated in the USA results from coal combustion, largely in boilers that are 20 – 50 years old [2]. Other sources include nuclear power, hydroelectric power, a growing commitment to natural gas, and minor uses of biomass, wind, and other renewable energy sources.
The average coal-fired boiler used for electricity generation exceeds 30 years of age. Fewer and fewer boilers have been installed since 1975. At the same time, however, capacities of coal-fired boilers have increased over time. By 1971 – 1975, for example, the average electric boiler being installed supported a turbine-generator producing nearly 600 MW_e. By 1990, the average new electric utility boiler supported a turbine-generator producing about 700 MW_e. In order to support growing demand for electricity, however, older boilers have been kept in service by life extension practices. Consequently there remains a very substantial population of utility boilers in the 150 – 350 MW_e capacity range [3].

The importance of electricity generation for fuel consumption trends cannot be overstated. Electrification has characterized all of the industrialized economies. Again using the USA as an example, fuel consumption for electricity generation has been the most prominent growth area within the energy sector as shown in Table 1.

Energy consumption in the "all other" category includes energy used in primary manufacturing—including such basic industries as petroleum refining, iron and steel, wood products, pulp and paper, aluminum, and chemicals. Like electricity generation, these industries do not rely upon oil and natural gas, necessarily, as their primary energy sources. Table 2 summarizes energy consumption patterns for selected industries as a function of fuel type or energy source.

The energy consumption patterns of electricity generation and the process industries show a significant potential for using opportunity fuels. Economic trends such as electricity generation deregulation, and increased global competition in the production of basic goods, provides significant incentive for owners of production facilities to seek maximum use of opportunity fuels as they favorably influence specific plant economics.

Environmental trends including stricter regulations regarding the formation and release of such pollutants as SO_2, NO_x and trace metals also promotes the use of opportunity fuels. Greenhouse gas emissions are also regulated in some nations, and are the subject of voluntary programs in other countries. These forces combine to increase the focus upon unconventional energy resources available to electricity generators and industrial establishments. The data in Table 2 demonstrate that, in reality, opportunity fuels are already penetrating several industrial sectors. They are increasingly economically significant for stationary energy users.

Table 1. Energy Consumption in the USA by Sector (values in %)

Year	Energy Sector			
	Electricity	**Transportation**	**Non-Fuel[1]**	**All-Other**
1950	13.3	24.5	2.9	59.3
1960	18.4	23.5	4.7	53.5
1970	24.2	23.7	4.7	47.4
1980	29.8	25.1	8.0	37.0
1990	35.0	26.8	7.1	31.1
2000	36.3	26.8	7.1	29.7

Note: [1]Non-fuel use includes production of chemicals, rubber, and plastics from oil and natural gas

Table 2. Distribution of Energy Consumption by Industry in the USA (values in %)

Fuel	Industrial Sector			
	Petroleum Refining	**Iron and Steel**	**Pulp and Paper**	**Wood Products**
Coal	0	44.8	12.0	0
Fuel Oil	1.2	2.1	7.0	6.0
Natural Gas	11.6	24.3	22.0	10.0
Electricity	5.4	24.7	8.0	14.0
Wood Residue	0	0	15.0	70.0
Pulping Liquor	0	0	36.0	0
Petroleum Coke	11.1	0	0	0
Refinery Gas	20.6	0	0	0
All Other	50.1	4.1	0	0

Source: [1]

The patterns of energy consumption in primary industries illustrate the potential for opportunity fuels. Note the number of opportunity fuels making significant contributions to the industrial sector: wood residues, spent pulping liquors, petroleum coke, refinery gas, and "all other". The category "all other" contains a wide array of energy sources including industrially owned hydroelectric facilities, locally generated materials in the agribusiness sector (e.g., rice hulls), landfill and wastewater treatment gases, combustible hazardous wastes used particularly in such industries as cement kilns, and many other specific examples of unusual combustible resources. With rising costs of premium fossil fuels, and increasing environmental requirements associated with the use of all fossil fuels, the utilization of these and other opportunity fuels continues to increase, and to gain increasing attention in the energy community.

Electricity generating utilities in the USA including Alliant Energy, Allegheny Energy Supply, TVA, Southern Company, Dynegy, and others are actively pursuing the use of opportunity fuels ranging from petroleum coke to various biomass fuels to waste coal products to a host of combustible resources unique to their locations [4 – 11]. Conferences ranging from the Coal Technology Association International Coal Utilization Conference in Clearwater, FL, the International Joint Power

Generation Conference of the American Society of Mechanical Engineers, the Electric Power Conference, Powergen, and many others are focusing increasing attention on this arena.

1.2. Towards a Definition of Opportunity Fuels

The term "opportunity fuels" is typically defined by listing combustible resources that are used within such a context. The common approach is "I know it when I see it." However a more universal approach also can be taken. Opportunity fuels can be defined as those combustible resources that are outside of the mainstream of fuels of commerce, but that can be used productively in the generation of electricity or the raising of process and space heat in industrial and commercial applications. Under certain circumstances, materials once defined as opportunity fuels can emerge as mainstream and highly popular energy resources. Letheby [4, 5], for example, contends that Powder River Basin coals originally were opportunity fuels that emerged as a dominant commercial energy source when industry learned how to handle and use these materials, and when regulatory and economic pressures provided a highly receptive marketplace.

Because opportunity fuels are outside the mainstream of fuels of commerce, the most common types are residues or low value products from other processes. These can include (not exhaustive) petroleum coke, sawdust, hogged wood waste (mixtures of sawdust, hogged bark, planer shavings, and other solid wood products residues), spent pulping liquor, rice hulls, oat hulls, edible pig lard from rendering plants, wheat straws and other straws from agricultural activities, unusable hays, peat, and a host of other solid materials [12, 13]. In addition to solid residues, blast furnace gas, coke oven gas, refinery off-gases, and like products also can be considered as opportunity fuels, used primarily on-site in process industries. Many of these fuels enter commerce to some extent, while others are strictly used on-site. Of this entire range of fuels, petroleum coke and wood wastes are the most widely used opportunity fuels, with the wood fuels alone supplying about 3 EJ (or quads) to the US economy and about 20 EJ worldwide [10].

A second category of opportunity fuels includes reclaimed and reprocessed wastes from energy industries and other industries. Such materials include anthracite culm, gob, slack, and other coal mining wastes. Fines from coal processing plants, including fines previously impounded in ponds, can be made into coal-water slurries and used as opportunity fuels. Certain industrial wastes also fit in this category including the byproduct aromatic carboxylic acid (BACA) burned by the Tennessee Valley Authority at its Colbert Fossil Plant.

A third category of opportunity fuels includes unconventional energy resources mined, extracted, or otherwise produced. Orimulsion™ fits this category. Orimulsion™ is used in several power generating stations throughout the world. In California, proposals have been made to burn oil soaked diatomite for energy production as well as recovery of diatomaceous earth; such material would serve as an opportunity fuel if utilized. Crops grown for energy purposes also could be considered in this category. At present, switchgrass is being grown experimentally for use at the Ottumwa Generating Station in Chillicothe, IA, and at Plant Gadsden in Gadsden, AL. Miscanthus, a fast-growing grass product, is being grown by Dynegy. Willow and hybrid poplar are also being grown experimentally as energy crops. Willow is being grown and used in the United Kingdom experimentally as well; the UK has an extensive program to use biomass fuels. Such crops represent another approach to producing an unconventional energy product directly.

Post-consumer materials provide the final category of opportunity fuels. Representative post-consumer materials include tire-derived fuel (TDF), commonly used in cyclone boilers, stoker boilers, and cement kilns. Other post-consumer materials used as opportunity fuels include waste oil and re-refined oil, wastewater treatment gas, landfill gas, paper-derived fuel (PDF), plastics-derived fuel, refuse-derived fuel (RDF), sewage sludge, wastewater treatment gas, and selected hazardous wastes burned in industrial boilers and kilns.

The range of opportunity fuels, depending upon source, is as broad as local opportunities and permitting agencies allow. The sources are based upon the principles of recycle and reuse, rather than disposal; the sources are also based upon winning the maximum and complete value from each raw material extracted from the earth's store of natural resources. Some of these opportunity fuels are quite new. Other opportunity fuels such as petroleum coke, wood waste, blast furnace gas, coke breeze, and refinery off-gas have a long tradition of use.

1.3. Typical Opportunity Fuel Applications

Opportunity fuels are used in a wide variety of applications, including electric utility boilers, industrial boilers and kilns, and institutional energy systems (e.g., prisons, hospitals, colleges and universities). Frequently these fuels are used to supplement traditional fossil energy sources such as coal; alternatively they may be used as the sole fuel for a given boiler or kiln.

1.3.1. The Use of Opportunity Fuels in Electric Utility Boilers

The use of opportunity fuels in electric utility applications is particularly significant, given that electricity generation is the most significant growth sector in the energy arena, as shown in Figure 1.2.

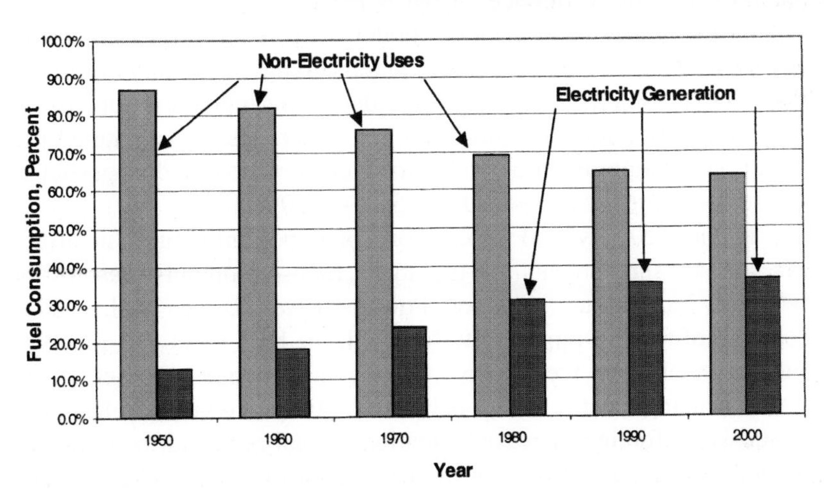

Figure 1.2. Fuel Consumption for Electricity Generation and Non-Electricity Uses in the USA.
(Source: [1])

Fuel consumption for all other applications—in aggregate—has remained constant while fuel consumption for electricity generation has continued to grow. As shown previously in Table 1, fuel consumption in the transportation sector has risen modestly, offsetting decreases in fuel consumption by the manufacturing and commercial sectors. Further, given the role that coal plays in electricity generation now, and the fact that the US Energy Information Agency (USEIA) projects coal to supply over half of the electricity generated in the US for the next 20 years [1], opportunity fuels have a significant potential in this arena.

Despite the recent growth in natural gas-fired combined cycle combustion turbines (CCCT's), wall-fired pulverized coal (PC) boilers, tangentially-fired PC boilers, and cyclone boilers are the dominant sources of electricity generation. In both types of PC boilers, coals or appropriate opportunity fuels are crushed, and then pulverized to <200 mesh, and then fired in the primary furnace section of the boiler. Typical firing conditions include stoichiometric ratios on the order of 1.15 – 1.20, and combustion temperatures on the order of 1480°C (2700°F). In cyclone boilers coal is crushed to <6.25 mm (<¼") and then fired in large cyclone barrels attached to the front (and potentially rear) wall of the cyclone boiler primary furnace. Coal is not pulverized for these units. Stoichiometric ratios for cyclone boilers are comparable to those associated with PC boilers. Combustion temperatures typically are in the 1650 – 1950°C (3000 – 3550°F) range depending upon the fuel being burned. Recently circulating fluidized bed boilers also have entered the electricity generating arena as well, firing crushed coal and opportunity fuels at temperatures of 815 – 900°C (1550 – 1750°F).

In the USA, typical applications of opportunity fuels in utility boilers are as supplements to the main fuel—commonly eastern or Midwestern bituminous coal, or Powder River Basin coal. This approach—cofiring an opportunity fuel with the base fuel—has received increasing attention in recent years, largely in response to the changing forces influencing electric utility behavior. Utility deregulation has been the first driver, forcing increased attention on fuel cost, and the total cost associated with using one or another coal or coal substitute. Deregulation has forced utilities to broaden their product slate to include "green power" or environmentally friendly power. This power, typically sold at a premium, results from generation of electricity from renewable resources in an environmentally sensitive manner. Deregulation also has come with a variety of regulations promoting the use of certain alternative energy sources including biomass, landfill and wastewater treatment gases, and—in some cases—municipal solid waste (MSW). These regulations typically take the form of Portfolio Standards, and numerous states such as New Jersey, Wisconsin, Texas, and Arizona have such laws mandating that utilities generate and/or sell a specific portion of their power based upon renewable technologies including wind, solar energy and—for some states—biomass. States such as Maryland offer production tax credits for power generated from biomass. Federal programs funded by USDOE during the late 1990's and the 2000 – 2002 time frame promoted the use of biomass opportunity fuels to address the CO_2 issue. These programs included biomass cofiring at such locations as Willow Island, WV (see Figure 1.3) where wood waste and TDF are cofired with coal.

Figure 1.3. Receiving biomass for cofiring at Willow Island Generating Station of Allegheny Energy Supply Co., LLC.

With the advent of deregulation came the concept of customer choice in selecting an electricity provider. Customer service, consequently, became a significant issue promoting the use of opportunity fuels. Customer service—accepting residues and low value products from electricity customers as fuel—permitted utilities to maintain their highly valuable industrial customer base. This factor has led to the firing of sawdust in some cases; it has led to the firing of a host of industrial byproducts as opportunity fuels.

Certain environmental regulations also have promoted the cofiring of opportunity fuels such as wood waste, herbaceous materials, landfill gas, wastewater treatment gas, sewage sludge, and the like. The use of opportunity fuels to reduce NO_x emissions, trace metal emissions, and like pollutants has been a powerful driving force encouraging their use. In Europe, response to the potential for global climate change and global warming has resulted in regulations mandating the use of biomass opportunity fuels. In Denmark, for example, the use of straw as a fuel is mandated by law. Other countries such as the UK have similar programs. In the USA, certain states such as Massachusetts tax CO_2 emissions; voluntary programs at the Federal level are used rather than regulatory mandates. Under the voluntary programs electric utilities such as Allegheny Energy Supply Co., LLC. have significantly reduced fossil CO_2 emissions. Allegheny has reduced its fossil CO_2 emissions by >1.7 million tonnes since measurement and voluntary programs were initiated. Other utilities have taken significant measures as well. Fossil CO_2 mitigation was one of the driving forces behind USDOE research concerning biomass cofiring at the National Energy Technology Laboratory (NETL) and the National Renewable Energy Laboratory (NREL), leading to programs such as the cofiring at Willow Island.

Environmental regulations governing landfills and waste disposal have promoted the use of TDF, PDF, and other opportunity fuels based upon post-consumer wastes. At the same time environmental regulations and regulatory interpretations have been less favorable to opportunity fuels such as petroleum coke and Orimulsion™. Regulatory programs such as New Source Review (NSR) have, in several cases, created particular difficulties in partial fuel switches to many of the opportunity fuels. The Boiler and Industrial Furnace (BIF) Rules have brought order to the firing of hazardous wastes in boilers; however they have increased the regulatory requirements associated with such practices. Driving forces to use opportunity fuels in electric utility boilers, then, range from reducing total fuel costs to improving boiler performance to addressing customer needs to achieving environmental gains. Consequently many electricity generating utilities in the US and in Europe regularly investigate, and commonly fire, opportunity fuels.

With all of these pressures, Letheby [4] concludes that the baseload coal-fired power plant is among the most promising markets for opportunity fuels. These plants can utilize up to 400,000 tonnes/yr at blend rates up to 20 percent—which is the common limit for using opportunity fuels in existing PC and cyclone boilers. At 20 percent (heat input basis) the opportunity fuels can significantly impact fuel cost, environmental performance, and other technical and economic parameters of such units.

1.3.2. Cofiring Opportunity Fuels in Process Industries and Independent Power Producers

Numerous process industries consume their own residuals as opportunity fuels. Pulp and paper mills consume bark and wood waste, spent pulping liquors, and pulp mill sludge as fuel. Spent pulping liquors are fired first for chemical recovery and for the generation of process steam or for power generation or both. Some consume post-consumer waste paper in the form of recycle sludge. Steel mills consume blast furnace gas, coke oven gas, and other internally generated combustibles along with fossil fuels in order to be most cost effective. Petroleum refineries consume petroleum coke and refining off-gas in the production of refined oil products.

Many additional process industries consume opportunity fuels in the production of product. Cement kilns commonly fire TDF and light liquid hazardous wastes in the production of clinker. Expanded aggregate kilns and other ore roasting kilns utilize the same hazardous wastes in their operations. These companies use opportunity fuels strictly as a means to reduce production costs in highly competitive environments.

Independent power producers (IPP's), created initially by the passage of the Public Utilities Regulatory Policies Act (PURPA), are another group dependent heavily upon opportunity fuels. Unlike electric utilities and process industries, however, IPP's tend to use opportunity fuels as 100 percent of the feed to any boiler. In eastern Pennsylvania,

IPP's have constructed numerous circulating fluidized bed (CFB) boilers fired with anthracite culm—a waste generated by the mining of hard coal. In California, IPP's built over 800 MW_e of capacity firing vineyard prunings, orchard prunings, and a wide variety of agricultural materials and wood wastes. While some of this capacity has been mothballed, over 500 MW_e is currently in operation. In Modesto, CA and elsewhere, IPP's have built generating stations fired totally with waste tires.

In Europe, utilities and industries burn biomass opportunity fuels extensively in operations similar to the IPP units. Commonly, the Northern European facilities generate both electricity and district heat in combined heat and power (CHP) facilities. These facilities are fired with wood waste, agricultural wastes, peat, municipal solid waste, spent pulping liquor, and a host of other related biomass products.

Despite the use of opportunity fuels in stand-alone operations, however, the most common use—and the most promising use—remains in cofiring applications [11]. This results from the number of coal-fired boilers in the existing fleet of units; the potential benefits of fuel costs, emissions management, and other technical improvements resulting from cofiring wood waste; and the ability to manage the risk of using opportunity fuels by firing them in situations where they can be removed if problems occur.

1.4. Issues Associated with Opportunity Fuel Utilization

The use of opportunity fuels must be considered within the framework of potential technical and economic impacts. The basic issues associated with using these energy sources include sourcing, or obtaining the fuels in a consistent and cost-effective manner. These fuels commonly are bought and sold outside the normal commercial fuel channels. Issues to be considered include such technical concerns as the impacts of opportunity fuels on the materials handling system—whether the opportunity fuel can be received and fed to the boiler reasonably—and on boiler performance considerations such as system capacity, efficiency, combustion rates and processes, and emissions formation. The issues also include the impacts of using opportunity fuels on total system economics including fuel cost, monetized emissions costs (e.g., potential SO_2 credits), and impacts on other cost areas such as ash sales or ash disposal.

1.4.1. The Influence of Opportunity Fuels on Combustion Processes

Opportunity fuels have the potential to influence the processes of solids combustion significantly, depending upon opportunity fuel. For example the petroleum cokes are typically low in volatile matter, moisture, and ash; at the same time they are high in calorific value and sulfur content. The waste coal products such as coal-water slurry are high in both moisture and ash. The biomass fuels are very high in volatile matter and moisture content, virtually free of any sulfur content, and can be quite low in ash content depending upon biomass resource. In order to provide a framework for understanding the technical influences of opportunity fuels it is important to review the basic mechanisms of solid fuels combustion—the mechanisms that can be significantly impacted by introducing new and dissimilar fuels to the boiler. The review presented below is not intended to discuss the opportunity fuels per se; rather, it is intended to provide a framework for combustion analyses presented in subsequent chapters.

1.4.1.1. The Overall Combustion Mechanism.

The general combustion mechanism, summarized in Figure 1.4, has been detailed in numerous publications [e.g., 14, 15] and involves the following sequential stages: particle heating and drying, pyrolysis and devolatilization, volatile oxidation, and char oxidation. For solid fuels, pyrolysis occurs in two stages. During the first stage bridges between aromatic clusters break, forming smaller molecules. In the second stage, functionalities and atoms are stripped from the core clusters, resulting in the formation of volatiles and tars. The volatiles evolve both as tars or condensable volatiles, and light gaseous molecules and fragments. The tars then further crack into gaseous molecules and char [14].

The critical issues associated with manipulating this mechanism include volatile yield in the combustor (the distribution between volatile matter and char), devolatilization kinetics, and char oxidation kinetics. In the management of emissions formation (e.g., NO_x emissions), manipulation of specific mechanisms becomes important. Fuel particle size, heating rate, and combustor temperature influence the proportional distribution between volatile matter and char. The chemical structure of the fuel—various coals, coal waste, petroleum coke, wood waste,

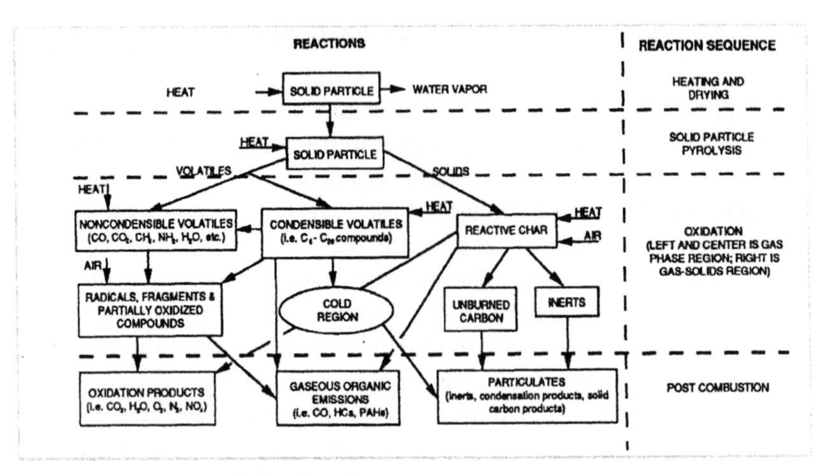

Figure 1.4. Processes of Combustion

switchgrass, tire-derived fuel—has a dominant influence in determining maximum volatile yield and pyrolysis kinetics.

Research by The Energy Institute of Pennsylvania State University and Foster Wheeler Power Group Inc., conducted for the Electric Power Research Institute, determined maximum volatile yield and devolatilization kinetics, along with fixed carbon/volatile matter ratios for a suite of fuels ranging from fresh sawdust and switchgrass through North Dakota lignite, Black Thunder and Cordero Rojo Powder River Basin coals, to Illinois #6 coal, Pittsburgh #8 coal, and petroleum coke. Maximum volatile yield and pyrolysis kinetics were determined using a drop tube reactor (DTR) at temperatures ranging from 400°C to 1700°C (752°F – 3092°F). Chemical structure was determined using Carbon 13 Nuclear Magnetic Resonance (^{13}C NMR). This research has been reported in numerous publications and papers [10, 11, 16-18].

Maximum volatile yield as a function of structure conforms to the following equation:

$$MVY(\%) = [1.011 - 0.314(A) - 0.01(AC/Cl)] \times 100 \qquad [1\text{-}1]$$

Where MVY(%) is percentage maximum volatile yield of the dry fuel, A is aromaticity (percentage), and AC/Cl is the average number of aromatic carbons per cluster. The r^2 for this equation is 0.955 and the probability that the equation, and the various terms, occurred randomly is as follows:

overall equation, 1.89×10^{-5}; intercept, 3.93×10^{-9}; A, 0.003; and AC/Cl, 0.033. Aromaticity, the number of aromatic carbon atoms divided by the total number of carbon atoms in a sample x 100, is the dominant driver with the size of the aromatic clusters being of secondary significance. Typical maximum volatile yields for bituminous coals are on the order of 55 – 65 percent, and maximum volatile yields for lower rank coals and lignites are on the order of 65 – 75 percent.

Devolatilization kinetics, or fuel reactivity, determined by DTR experiments, was based upon bulk furnace temperatures rather than particle temperatures; as a consequence these kinetics are not directly appropriate for computational fluid dynamics. However they are a highly useful measure of fuel reactivity [17-18]. Pyrolysis kinetics can be calculated as a series of Arrhenius equations. The typical pyrolysis reactivity for a PRB coal is shown as an Arrhenius plot in Figure 1.5.

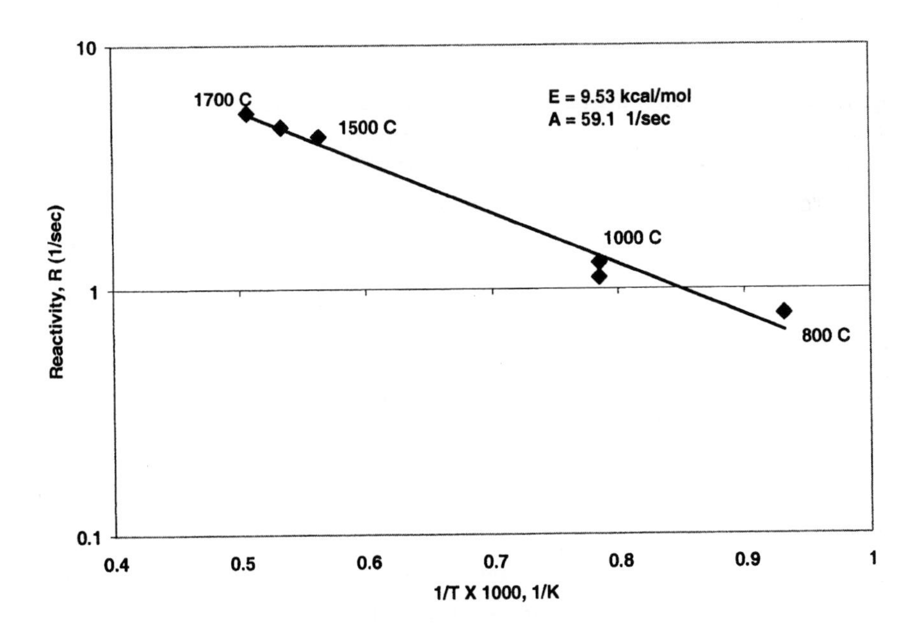

Figure 1.5. Devolatilization Reactivity for Black Thunder Coal
Source: [17]

Over the range of fuels tested, the reactivities then could be correlated to structure. Activation energies in the Arrhenius equations were largely a function of structure as shown by equation 1-2 [17]:

$$E_{act} = [1.97 + 17.22(A) - 0.20(AC/Cl)] \times 4.19 \qquad [1\text{-}2]$$

Where E_{act} is activation energy in kJ/mol. The r^2 for this equation is 0.867 and the probability that it occurs as a random event is 8.64×10^{-4}. The driving force in this equation is aromaticity (A), as can be seen from Figure 1.6.

From the burner design perspective, this fuel reactivity is typically expressed as the fixed carbon/volatile matter (FC/VM) ratio derived from the proximate analysis. The influence of aromaticity on that ratio can be observed from equation 1-3:

$$FC/VM = 0.1587 \times e^{3.1421(A)} \qquad [1\text{-}3]$$

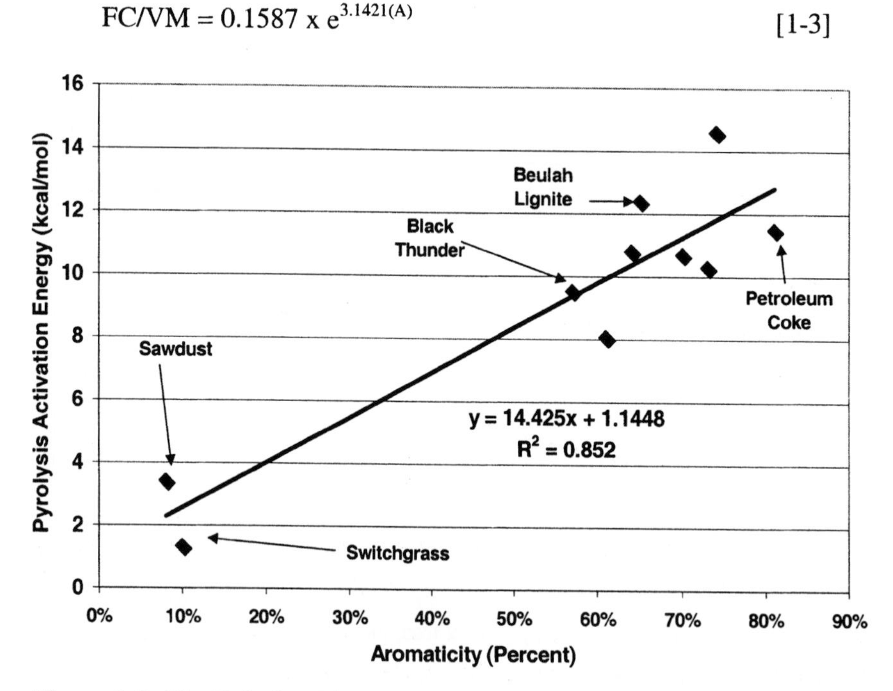

Figure 1.6. The Relationship Between Aromaticity and Fuel Reactivity or Devolatilization Activation Energy

The r^2 for equation [3] is 0.955 over the complete range of fuels tested. Figure 1.7 depicts this relationship.

Evaluated on this basis, the opportunity fuels have a significant potential to impact the reactivity of a fuel introduced into a boiler. Depending upon the base coal, the percentage opportunity fuel being fired, and the reactivity of the opportunity fuel—from the highly reactive biomass fuels to the far less reactive petroleum cokes—there can be a significant impact on the combustion process as a consequence of using these alternative energy resources. Such impacts are evaluated in subsequent chapters.

1.4.1.2. NO_x Formation and Control Mechanisms

Conventionally NO_x formation is divided into thermal NOx and fuel NO_x, with the former coming from either the Zeldovich mechanism

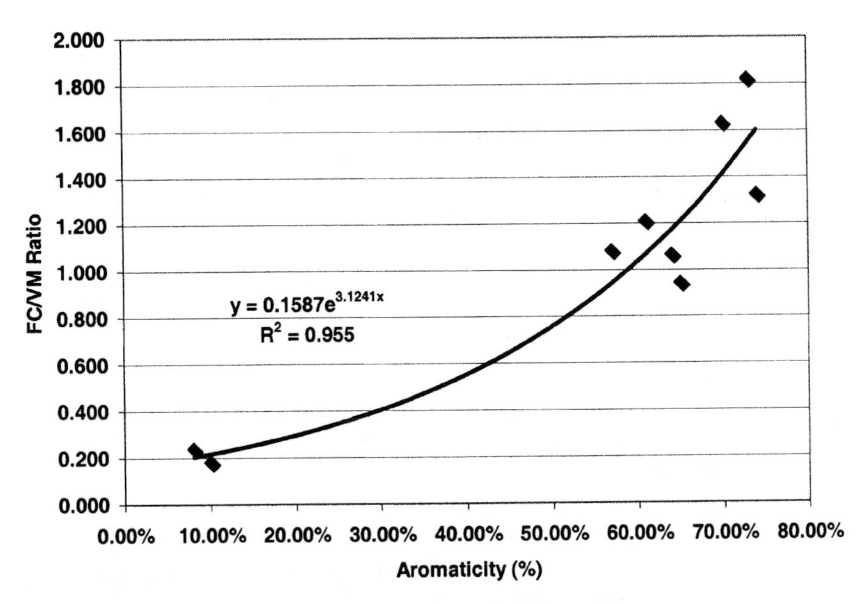

Figure 1.7. The Influence of Aromaticity on the Fixed Carbon/Volatile Matter Ratio Determined from the Proximate Analysis

or the "prompt NO" mechanism elucidated by Fenimore and others [14, 19]. Fuel NOx results from oxidation of nitrogen found in the fuels being burned. That nitrogen exists in a reduced condition, either in pyridine, pyrrole, substituted pyrrole, or amine form depending upon fuel being burned. Most NO_x formed results from the oxidation of fuel nitrogen, particularly with current combustion technology. Consequently much of the technology developed to manage this pollutant is in the form of combustion modifications: low NO_x burners, separated overfire air in its many configurations, reburn systems, and flue gas recirculation. Figure 1.8 summarizes the potential fates of fuel nitrogen in combustion systems.

Baxter et. al. [20] have demonstrated that a critical phenomenon is the rate of fuel nitrogen evolution in volatile form, relative to the rate of total nitrogen evolution. This research shows that volatile nitrogen evolution from coal typically lags behind total volatile evolution when comparing the percentage of the nitrogen volatilized to the percentage of total fuel volatilized. Different coals have different volatile evolution pathways, however [20, 21].

Conventional fuel characterization methods do not always provide sufficient information to evaluate opportunity fuels—particularly with respect to nitrogen evolution and NO_x formation. Consequently additional techniques for characterization and analysis have been developed to analyze the fate of fuel nitrogen and its evolution in volatile

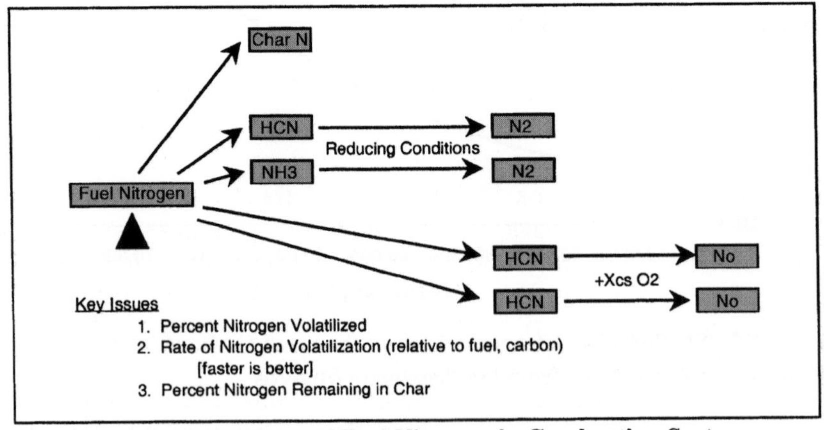

Figure 1.8. Potential Fates of Fuel Nitrogen in Combustion Systems

or char form during pyrolysis reactions. Research at The Energy Institute of Pennsylvania State University has shown that various opportunity fuels have nitrogen evolution patterns that are markedly different from those associated with coals [17]. In some cases—notably fresh sawdust—the nitrogen evolves more rapidly than the total mass of volatiles, a result consistent with previous research by Tillman and Smith [22]. In other cases such as petroleum coke, volatile nitrogen evolution and total volatile evolution proceed at about the same rate. Further, the research by Johnson et. al. [17] shows marked differences between fuels in the amount of fuel nitrogen remaining in the char. Nitrogen retained in the char evolves more slowly. Consequently, like the volatile nitrogen that evolves slowly, it is most likely to be reacted in fuel-lean conditions rather than substoichiometric conditions. These patterns of nitrogen evolution have significant impacts on NO_x formation. As can be readily inferred, cofiring opportunity fuels with distinct and different nitrogen evolution patterns—and total volatile evolution patterns—can significantly impact NO_x formation. This parameter, too, must be evaluated in the subsequent chapters.

1.4.1.3. Slagging and Fouling Issues

Slagging and fouling issues also must be considered when adding opportunity fuels to the total mass of fuel entering a boiler or combustion system. The ash chemistries associated with many of the opportunity fuels are fundamentally different from the ash chemistries associated with most coals being burned today.

Bulk ash chemistries determining the concentrations of SiO_2, Al_2O_3, TiO_2, Fe_2O_3, CaO, MgO, Na_2O, K_2O, P_2O_5, SO_3, provide the first level of analysis, and determining slagging and fouling indices according to the Babcock & Wilcox [23] formulae offers the first indication of potential issues. Alternatively, Miles et. al. [24] have proposed a deposition index utilizing Na_2O and K_2O as primary indicators. In this index, 0.17 kg/GJ (<0.4 lb/10^6 Btu) Na_2O and K_2O indicates slight potential for slagging and fouling, while 0.34 kg/GJ (>0.8 lb/10^6 Btu) indicates high potential for slagging and fouling.

More recently, research by Miller et. al. [12, 25] has indicated that slagging and fouling issues associated with blends are more complex. Using FactSage thermodynamic modeling, Miller et. al. [25] has shown the ability to evaluate fuel blends with respect to slag formation and slag viscosity. This research has shown that, in cofiring situations, some opportunity fuels such as cattle manure have the potential to form eutectics that significantly reduce ash fusion temperatures and ash viscosities. Other opportunity fuels such as wood waste have significantly less potential for creating such problems. Research by Miller et. al. [12] and by Baxter et. al. [26, 27] have shown that chemical fractionation provides additional critical insights into the behavior of inorganic constituents in fuels. The chemical fractionation technique was originally developed by Miller and Given at The Pennsylvania State University in the 1970's as a way to evaluate biomass ash. It was subsequently applied to western coals by the University of North Dakota Energy & Environmental Research Center, by Foster Wheeler Development Corporation, and by several commercial laboratories such as Hazen Research of Golden, CO. Its use has been amplified through application in the past decade.

Chemical fractionation—successive leaching of the fuel in demineralized water, ammonium acetate, and hydrochloric acid, provides critical insights into the reactivity of ash. It is particularly useful in evaluating low rank coals, biomass fuels, and related combustible resources; it can be used to evaluate bituminous coals as well, and almost uniformly shows the higher propensity of low rank fuels to create problems.

The issues of slagging and fouling also include corrosion, and in this arena the opportunity fuels can play a role in influencing the performance of boilers and combustion systems. Corrosion mechanisms are many and complex; several are caused by chlorine. Chlorine is a constituent in several coals; it is also a common constituent in agricultural biomass, and wood waste if the wood has been stored in salt water ponds. Baxter et. al. [28] has shown that the KCl is particularly aggressive, and that this can be introduced by agricultural biomass. Vanadium also can be an agent causing corrosion, particularly if the vanadium oxidizes completely to V_2O_5. Alternatively, since vanadium is the base for many of the selective catalytic reduction catalysts, the presence of vanadium in petroleum coke or other opportunity fuels potentially can assist in the performance of the SCR. Similarly, for boilers equipped with "hot side" electrostatic precipitators, the presence of sodium in the inorganic constituents of low rank opportunity fuels can improve the performance of the pollution control device. The presence of alkali and alkali earth elements in biomass fuels has been used to reduce the T_{250} temperatures of fuel being fed to cyclone boilers—with beneficial results.

The major inorganic constituents in opportunity fuels, then, must be evaluated along with the basic combustion processes. Opportunity fuels can cause significant problems, or improve operations, depending upon the specific fuel and the specific situation. This issue merits exploration as a function of each type of opportunity fuel.

1.4.1.4. Trace Metal Considerations

Trace metals, particularly mercury but also including such metals as (not exhaustive) arsenic, barium, beryllium, cadmium, chromium, copper, lead, nickel, and zinc, are of final consideration with respect to the technical aspects of using opportunity fuels either in cofiring situations or in stand-alone boilers. Mercury is well on its way to being regulated. Arsenic has the potential to poison SCR catalysts. All such metals are governed by the Toxic Release Inventory (TRI) processes of the USEPA.

Trace metal concentrations vary significantly as a function of coal or oil being burned. At the same time trace metal concentrations in opportunity fuels vary dramatically. Petroleum coke, for example, contains very low concentrations of mercury and arsenic, but high concentrations of vanadium and nickel [29]. Tire-derived fuel has high concentrations of zinc, as a function of using zinc to accelerate the vulcanizing process [30, 31]. Fresh sawdust contains very little mercury [6, 7], however wood typically has high concentrations of barium [15]. All biomass fuels contain varying concentrations of trace metals depending upon the soils where they are grown [15]. The use of hazardous wastes as opportunity fuels necessarily involves careful consideration of this issue.

Trace metal issues are of particular importance when firing waste oils and hazardous wastes. These fuels come under the Boiler and Industrial Furnace (BIF) Rules, and as such merit additional scrutiny in this arena. Many of the waste products—from municipal wastes to hazardous wastes—can contain measurable concentrations of trace metals requiring attention. Further, utilities must report metals such as mercury under the Toxic Release Inventory (TRI) regulations of the US

Environmental Protection Agency (USEPA)—again bringing attention to the trace metals considerations when firing opportunity fuels.

The analysis of opportunity fuels in subsequent chapters, then, must consider trace metal concentrations in the various types of material being evaluated. Of particular concern are those metals being scrutinized by environmental agencies.

1.4.2. Economic Considerations with Opportunity Fuels

The economic considerations associated with using opportunity fuels have been well identified by Letheby [4, 5] and Williams [32]. These issues extend beyond the cost of the fuel in \$/GJ (\$/10^6 Btu). Among the economic issues is fuel transportation cost. Many opportunity fuels must be located near the power plant or energy user, due to inordinate impacts of transportation on final fuel costs. In addition, economic impacts include the monetary consequences of using the fuel with respect to boiler performance: capacity, efficiency, and operability. They also include the impact of the fuel on emissions and emissions credits; reductions in SO_2 and NO_x emissions currently have significant economic values in the US.

From an economic perspective, opportunity fuels also must be considered with particular attention to power plant or fuel user selection. Those utilities or process industries seeking to use petroleum coke, for example, must select boilers equipped with acid gas removal systems. Such systems may be scrubbers; alternatively they may be fluidized bed boiler systems where limestone is added for sulfur capture.

Beyond fuel cost, the economic issues include capital costs and operating costs for installing a system. These costs are particularly significant with respect to fuel preparation. Certain petroleum cokes, for example, may have a low Hardgrove Grindability Index and may be difficult to grind. Biomass fuels require hammer mills and other grinding equipment capable of cutting or shredding the fuel due to the fibrous nature of biomass fuels. Biomass fuels are also lower in bulk density than coal or other opportunity fuels such as petroleum coke; this, also, impacts capital and operating costs when considered on a \$/GJ basis. In an era of downsized power plant operations and maintenance staffs, the operating and maintenance costs are of particular importance. Operating costs can include additional reagent for scrubbing flue gas laden with sulfur dioxide

(SO_2), oxides of nitrogen (NO_x), or particulates if such problems are introduced by opportunity fuels. Such costs must be weighed when considering one or another such energy source.

When these economic issues are properly addressed, however, opportunity fuels can be of significant benefit to both older generating stations and the most modern of technologies. Opportunity fuels have been successfully fired in cyclone boilers installed in the late 1950's and 1960's. They have also been successful in the most modern of power generating stations. The Polk County, FL 250 MWe integrated gasification-combined cycle (IGCC) generating station shown in Figure 1.9, for example, routinely fires a blend with >50 percent petroleum coke, and it has cofired ground eucalyptus wood with coal and petroleum coke as well.

Opportunity fuels may have additional economic benefits—generating green or clean power by firing biomass is but one example. The use of waste paper or plastics, or hazardous wastes also can bring additional economic benefits if tipping fees can be charged. Firing wastewater treatment gas as a reburn fuel can reduce NO_x emissions if the system is appropriately designed and installed.

Of particular concern in assessing the economics is the impact of opportunity fuel utilization on waste disposal and permitting. For coal-fired power plants that sell flyash under ASTM Specification C-618, using opportunity fuels can compromise the ability to sell such material. This can swing the economics of use by >$1 million/year in many cases—even with small boilers. The contemplation of opportunity fuels has raised permitting issues in several cases, and permitting issues have significant economic implications. Regulation is uneven across the USA, and the industrialized world, with respect to its application.

The use of biomass, TDF, and other similar materials can have non-monetized benefits in terms of public image and public acceptance. Alternatively, local groups have vigorously opposed the use of some opportunity fuels. Opportunity fuels also can be opposed by fuel yard workers, if those workers feel that they are exposed to health issues. Manures, for example, have to be handled separately from coal. Economic considerations tend to be incredibly site-specific. Opportunity fuel availability and cost is subject to local conditions of supply and demand; opportunity fuels frequently are bought and sold outside the normal commercial channels of the energy market. Capital costs of

Figure 1.9. The Gasifier at the Polk County Generating Station, which fires a blend of petroleum coke and coal and which as cofired ground eucalyptus wood as well.

opportunity fuel systems depend significantly upon site conditions at a given power plant. These conditions include space available and its proximity to the power house, soil conditions at the site, local codes and ordinances, and the standards and preferences at the plant site. Capital costs also depend heavily upon design criteria: the extent of automation required, the extent to which plant labor can be available to operate a system, the design life, system availability and associated redundancy requirements, etc.

Because economic conditions are highly site specific, generalized comments will be made in subsequent chapters only as indicators of what can be done. Detailed economics cannot be presented without a high potential to be misleading. Case studies, however, are presented for each fuel selected in order to show the possibilities as well as the technical possibilities.

1.5. Approach of this Assessment

Given the wide array of possible opportunity fuels, and the myriad of issues that are associated with their use, this text takes a particular approach to evaluating these energy resources. The text recognizes that successful use of opportunity fuels requires successful integration of these energy resources into existing firing systems, should cofiring be the approach taken. Integration includes materials receiving, materials preparation, firing, ash handling, and pollution control.

Successful implementation of an opportunity fuel program must accomplish technical, environmental, and/or economic improvements. These can be increases in boiler capacity or efficiency, either through the use of a high calorific value fuel (petroleum coke, TDF) or by the design of the system used to introduce the fuel into the boiler (e.g., separate injection of biomass into a PC boiler can overcome pulverizer limitations on capacity). There can be reductions in emissions of SO_2, NO_x, CO and hydrocarbons, and mercury and other trace metals associated with the use of opportunity fuels. TDF fired with medium sulfur bituminous coal can have a significant benefit with respect to SO_2 emissions. Improperly used, however, opportunity fuels can increase certain emissions including opacity.

Opportunity fuels then, can create advantages either in cofiring or stand-alone operations. At the same time there are limitations to their usage either caused by local supply and demand or by the combustion system in which they will be used.

This text approaches opportunity fuels by selecting representatives of various categories of materials: petroleum coke; Orimulsion™; waste coal products and coal-water slurries made from coal pond fines; woody biomass fuels including both fresh wood wastes and urban wood wastes; agricultural biomass fuels including crops, crop wastes, and manures; tire-derived fuel and associated post-consumer waste-based fuels; and gaseous and liquid opportunity fuels ranging from coal-bed methane to landfill and wastewater treatment gases, liquid hazardous wastes fired in cement kilns and industrial boilers, and related products. Not all opportunity fuels can be covered, however the selection provides insights into all categories of opportunity fuels. Further it provides insights into the major opportunity fuels used or contemplated today.

The focus of the book is the technical characteristics of these opportunity fuels, recognizing that these are the driving forces ultimately influencing utilization economics. Economics are dealt with in a general sense, and through case studies. This provides a useful approach in evaluating the opportunities and limitations of these energy resources available to broaden the fuel base for any user.

1.6. References

1. US Energy Information Agency, 2003. Annual Energy Review 2001. USEIA, Washington, D.C.
2. Hughes, E. and D. Tillman. 2003. Biomass Energy Utilization. EPRI. Palo Alto, CA.
3. Bergesen, C. and J. Crass (eds). 1996. Power Plant Equipment Directory. Utility Data Institute, Washington, D.C.
4. Letheby, K. 2002. Utility Perspectives on Opportunity Fuels. Proc. 27th International Technical Conference on Coal Utilization and Fuel Systems. Clearwater, FL. March 4-7.
5. Letheby, K. 2002. Utility Utilization of Opportunity Fuels. Proc. International Joint Power Generation Conference. Phoenix, AZ. June 22-25.
6. Tillman, D., K. Payette, T. Banfield, and S. Plasynski. 2003. Cofiring Woody Biomass at Allegheny Energy: Results from Willow Island and Albright Generating Stations. Proc. 28th International Technical Conference on Coal Utilization and Fuel Systems. Clearwater, FL. March 10-13.
7. Payette, K., T. Banfield, T. Nutter, and D. Tillman. 2002. Emissions Management at Albright Generating Station Through Biomass Cofiring. Proc. 27th International Technical Conference on Coal Utilization and Fuel Systems. Clearwater, FL. March 4-7.
8. Harding, N.S. 2002. Co-Firing Tire-Derived Fuel with Coal. Proc. 27th International Technical Conference on Coal Utilization and Fuel Systems. Clearwater, FL. March 4-7.
9. Zemo, B., D. Boylan, and J Eastis. 2002. Experiences Co-firing Switchgrass at Alabama Power's Plant Gadsden. Proc. 27th International Technical Conference on Coal Utilization and Fuel Systems. Clearwater, FL. March 4-7.
10. Tillman, D.A. 2002. Cofiring Technology Review, Final Report. National Energy Technology Laboratory, US Department of Energy, Pittsburgh, PA.

11. Tillman, D.A. 2001. Final Report: EPRI-USDOE Cooperative Agreement: Cofiring Biomass With Coal. Contract No. DE-FC22-96PC96252. EPRI, Palo Alto, CA.
12. Miller, B.G., S.F. Miller, C. Jawdy, R. Cooper, D. Donovan, and J. Battista. 2000. Feasibility Analysis for Installing a Circulating Fluidized Bed Boiler for Cofiring Multiple Biofuels and Other Wastes at Penn State University: Second Quarterly Technical Progress Report. Work Performed Under Grant No. DE-FG26-00NT40809.
13. Miller, B.G., S.F. Miller, and A.W. Scaroni. 2002. Utilizing Agricultural By-products in Industrial Boilers: Penn State's Experience and Coal's Role in Providing Security for our Nation's Food Supply. Proc. 19th Annual International Pittsburgh Coal Conference, Pittsburgh, PA Sep 23-27.
14. Tillman, D.A. 1991. The Combustion of Solid Fuels and Wastes. Academic Press, San Diego.
15. Tillman, D.A. 01994. Trace Metals in Combustion Systems. Academic Press, San Diego.
16. Johnson, D.K. et. al. 2001. Characterizing Biomass Fuels for Cofiring Applications. Proc. Joint International Combustion Symposium. American Flame Research Committee. Kaui, Hawaii. Sep 9 – 12.
17. Johnson, D.K., D. Tillman, and B. Miller. 2003. Reactivity of Selected Opportunity Fuels: Measurements and Implications. Proc. Electric Power Conference. Houston, TX. March 3-7.
18. Tillman, D., B. Miller, and D. Johnson. 2003. Nitrogen Evolution from Biomass Fuels and Selected Coals. Proc. 28th International Technical Conference on Coal Utilization and Fuel Systems. Clearwater, FL. March 10-13.
19. Palmer, H. 1974. Equilibria and Chemical Kinetics in Flames. in Combustion Technology: Some Modern Developments (H. Palmer and J. Beer, eds). Academic Press, New York.
20. Baxter, L.L., R.E. Mitchell, T.H. Fletcher, and R.H. Hurt. 1996. Nitrogen Release during Coal Combustion. ENERGY & FUELS. 10(1): 188-196.
21. Tillman, D.A. 2003. Biomass and Coal Characteristics: Implications for Cofiring. International Conference on Co-Utilization of Domestic Fuels. Gainsville, FL. Feb. 5-6.
22. Tillman, D.A. and W.R. Smith. 1982. The evolution of nitrogen volatiles from Red Alder bark. Proc. Forest Products Research Society, St. Paul, MN. June 23-25.

23. Stultz, S.C. and J.B. Kitto (eds). 1992. Steam: Its Generation and Use. 40th Ed. Babcock & Wilcox. Barberton, OH.
24. Miles, T.R., T.R. Miles Jr., L.L. Baxter, B.M. Jenkins, and L.L. Oden. 1993. Alkali Slagging Problems with Biomass Fuels. Proc. First Biomass Conference of the Americas. Burlington, VT. Aug 30 – Sep 2. pp. 406 – 421.
25. Miller, S.F., B.G. Miller, and D.A. Tillman. 2002. The Propensity of Liquid Phases forming During Coal-Opportunity Fuel (Biomass) Cofiring as a Function of Ash Chemistry and Temperature. Proc. 27th International Technical Conference on Coal Utilization and Fuel Systems. Clearwater, FL. March 4-7.
26. Baxter, L.L. 2000. Ash Deposit Formation and Deposit Properties: Final Report. SAND2000-8253. Sandia National Laboratories. Livermore, CA.
27. Baxter, L.L. et. al. 1996b. The Behavior of Inorganic Material in Biomass-Fired Power Boilers—Field and Laboratory Experiences: Vol I and II of Alkali Deposits Found in Biomass Power Plants. SAND96-8225 Volume 2 and NREL/TP-433-8142.
28. Baxter, L.L. 2003. Biomass Combustion and Cofiring Issues Overview: Alkali Deposits, Flyash, NOx/SCR Impacts. International Conference on Co-Utilization of Domestic Fuels. Gainsville, FL. Feb. 5-6.
29. Hus, P.J. and D.A. Tillman. 2000. Cofiring multiple opportunity fuels with coal at Bailly Generating Station. *Biomass and Bioenergy*. 19(6): 385-394.
30. Harding, N. S., 2002. "Cofiring Tire-Derived Fuel With Coal," 27th International Technical Conference on Coal Utilization and Fuel Systems, Clearwater, FL.
31. McGrath, J.F. 1985. "Elastomers, Synthetic" in Concise Encyclopedia of Chemical Technology, Kirk-Othmer, eds., John Wiley & Sons, Inc., Pp. 391.
32. Williams, C. 2003. Utility Perspectives on Opportunity Fuels. Proc. Electric Power Conference. Houston, TX. March 3-7.

CHAPTER 2: PETROLEUM COKE AND PETROLEUM-BASED PRODUCTS AS OPPORTUNITY FUELS

2.1. Introduction

Petroleum coke is a solid byproduct of petroleum refining, useful in the production of electrodes used as carbon anodes for the aluminum industry, graphite electrodes for steel making, as fuel in the firing of solid fuel boilers used to generate electricity, and as fuel for cement kilns. Currently North America produces a very high percentage of all petroleum coke, and coking capacity continues to increase. Refineries in Latin America are also increasing petroleum coke production capacity [1]. Petroleum coke is of increasing importance in the cement industry [1], and is growing in use within the electricity generating and industrial boiler communities as well. Further, as utilities shift their focus towards lower cost generation including solid fuel supercritical boilers [2] and continue to investigate integrated gasification-combined cycle combustion turbine (IGCC) generating stations, petroleum coke will continue to grow in importance as a solid fuel. Petroleum coke is the about 20 percent of the fuel used by JEA at its two 660 MW_e wall-fired boilers at the St. Johns River Power Park and at its new circulating fluidized bed (CFB) boilers; petroleum coke also provides over half of the fuel for the Polk County IGCC of Tampa Electric Company. Already, in the USA, over 1.5 million tonnes (1.68 x 10^6 tons) of petroleum coke are used by major utilities as is shown in Table 2.1. The primary reason is shown in Table 2.1—the low cost of petroleum coke compared to coal.

Table 2.1. Representative Petroleum Coke Consumption by US Electric Utilities in the Year 2000

Utility	Use (tonnes)	Use (tons)	Delivered cost ($/GJ)	Delivered cost ($/$10^6$ Btu)
Central Illinois Public Service	23400	26000	0.86	0.91
Jacksonville Electric Authority	400000	444000	0.58	0.61
Lakeland Dept of Water and Elect.	1800	2000	0.41	0.43
Manitowoc Public Utilities	32400	36000	0.45	0.47
Michigan South Central Power	1800	2000	1.02	1.07
Northern Indiana Public Service Co. (NiSources)	156800	174000	0.62	0.65
Northern States Power	198200	220000	0.31	0.33
Ohio Edison Co	7200	8000	0.70	0.74
Owensboro, City of	8100	9000	0.51	0.54
Pennsylvania Power	183000	203000	0.70	0.74
San Antonio, City of	8100	9000	0.40	0.42
Tampa Electric Co	190000	211000	0.48	0.51
Union Electric Co (AMEREN)	111700	124000	0.58	0.61
Wisconsin Electric Power Co.	132400	147000	0.66	0.70
Wisconsin Power & Light (Alliant Energy)	62200	69000	0.45	0.47
TOTAL (*)	1516200	1683000	0.55	0.58

(*) Note: Totals may not add due to rounding

2.1.1. Petroleum Coke Production Processes

Petroleum coke production processes are reviewed by Bryers [4, 5]; approaches include delayed coking and fluid coking. Delayed coking is a batch process where residual components of crude oil are heated to about 475 – 520°C (890 – 970°F) in a furnace and then confined in a coke drum for thermal cracking reactions. The products of the coking process are gas, gasoline, gas oil, and coke. In a typical delayed coking arrangement, two coke drums are used; one is being filled and reactions are proceeding while the produced coke is being removed by high-pressure water. The delayed coking process produces various types of petroleum coke including needle coke, granular coke, sponge coke, and shot coke. Sponge and shot cokes are commonly found in the fuel market. Sponge coke is highly porous and anisotropic in nature; shot coke, which appears like an assembly of small balls or beebees, also exhibits anisotropic characteristics, however it is less porous and much harder to crush. Shot coke often is the result of upset conditions in the delayed coker.

Fluid coke is produced in a fluidized bed reactor where the heavy oil feedstock is sprayed onto a bed of fluidized coke. The oil feedstock is cracked by steam introduced into the bottom of the fluidized bed reactor. Vapor product is drawn off the top of the reactor while the coke descends to the bottom of the reactor and is transported to a burner where a portion of the coke is burned to operate the process. Fluid coke reactors are operated at about 510 – 540°C (950 – 1000°F). Flexicoke is a variant on fluid coke, where a gasifier is added to the process to increase coke yields. Fluid coker installations tend to have yields that are lower than delayed coker installations, while flexicoker installations have yields that can be significantly greater than delayed coker installations. Fluid cokers produce layered and non-layered cokes. Both delayed and fluid coke installations produce amorphous, incipient, and mesophase cokes with the amorphous cokes having higher volatility and mesophase cokes having the lowest volatility.

Petroleum coking can be varied to impact petroleum coke fuel properties and quality, as has been shown by the development of the OptiFuel™ process of Environmental Energy Enterprises. This variability includes volatility, type of volatile matter, and Hardgrove Grindability Index.

2.1.2. Petroleum Coke Production and Utilization

Worldwide petroleum coke production now exceeds 127,600 tonne/day (141,800 ton/day) or 46,600,000 tonne/year (51,740,000 ton/year) and is continuously growing [1]. Jacobs Consultancy estimates that worldwide production of petroleum coke will exceed 58 million tonnes/year by the end of 2002 [6]. Jacobs reports an annual increase in petroleum coke supply, from 1996 to 2001, of 7.3 percent; and an increase in supply from 2001 to 2002 of 11.3 percent [6]. The distribution of petroleum coke production conforms to the following percentages: North America, 66.5 percent; Europe, 17 percent; Asia-Pacific, 9.5 percent; South America and the Caribbean, 4.5 percent; and the Middle East and Africa, 2.5 percent [7].

The growth of petroleum coke production goes beyond the growth in demand for petroleum products. Further, this increase in production has occurred despite depressed petroleum coke prices. In 2002 alone, 3 new cokers with 4.3 million tonne/yr of capacity were brought on line in Venezuela, Mexico, and St. Croix [8]. Prices vary from region to region, and are based upon coke quality measured by sulfur content. Regional price variations depend upon local supply-demand relationships. Petroleum coke is used throughout the world; nearly 2/3 of the petroleum coke produce in North America is exported to Japan, Europe, and other regions of the global economy [7].

The attractiveness of fuel grade petroleum coke extends beyond price characteristics. It is typically high in heat content (typically >32.68MJ/kg or 14,000 Btu/lb) and low in ash content (typically <1 percent ash). These characteristics favor its primary fuel use—as a fuel to be cofired with coal [7]. High sulfur concentrations and, for many petroleum cokes high vanadium and nickel concentrations, remain as the less desirable characteristics of this opportunity fuel. The high sulfur content limits the use of petroleum coke to boilers equipped with acid gas scrubbers—and with units where there is sufficient scrubbing capacity to handle the increased sulfur loading. Periodically, low Hardgrove Grindability Index (HGI) values also inhibit the use of some petroleum cokes. In reality, even within fuel grade petroleum coke there are substantial differences in fuel characteristics depending upon source of crude oil and coking method. These variations are explored in the subsequent discussions concerning petroleum coke fuel characteristics, and case studies of petroleum coke fuel use.

2.2. Fuel Characteristics of Petroleum Coke

Traditional analyses of fuels include proximate, ultimate, and ash elemental analysis along with calorific value and, increasingly, trace metal concentrations. These values are presented below; however fuel characterization increasingly needs to focus attention on additional measures of fuel structure, fuel volatility, and fates of certain elements such as fuel nitrogen.

2.2.1. Proximate and Ultimate Analysis of Petroleum Coke

There is a significant body of literature concerning the traditional characteristics of petroleum coke [e.g., 4, 5, 9-11], and the influence of coking processes and conditions on such properties [e.g., 4-5, 12]. Table 2 presents typical characteristics of various petroleum cokes as a function of coking method.

Note that the flexicoke has the lowest volatile content, and also the lowest calorific value. However all of the petroleum cokes are high in calorific value. On occasion, off-specification petroleum coke is also supplied to generating stations. A survey of tests conducted by others shows that petroleum cokes with 8 - 10 percent volatile matter are common; higher volatile matter (e.g., 10 – 14 percent volatiles) can also be encountered. In one case, the Bailly Generating Station cofiring tests, the petroleum coke had a volatile matter content of 14 percent [13-14]. The supply of petroleum coke with 12 – 15 percent volatile matter is not uncommon in the use of this opportunity fuel.

The data in Table 2.2 can be converted into certain empirical performance characteristics as shown in Table 2.3. In these empirical performance characteristics, hydrogen/carbon (H/C) and oxygen/carbon (O/C) atomic ratios from the ultimate analysis are used as volatility measures along with the volatile/fixed carbon (V/FC) ratio from the proximate analysis. The H/C and V/FC ratios are the dominant volatility measures; O/C is of less significance but is not insignificant in assessing fuel volatility.

Table 2.2. Typical Fuel Characteristics of Petroleum Coke[1]

Analysis	Fuel Type			
	Delayed coke (Sponge)	Shot coke	Fluid coke	Flexicoke
Proximate Analysis (wt %)				
Fixed Carbon	80.2	89.59	91.50	94.9
Volatiles	4.48	3.07	4.94	1.25
Ash	0.72	1.06	1.32	0.99
Moisture	7.60	6.29	2.24	2.86
Ultimate Analysis (wt %)				
Carbon	81.12	81.29	84.41	92.00
Hydrogen	3.60	3.17	2.12	0.30
Oxygen	0.04	0.93	0.82	0.00
Nitrogen	2.55	1.60	2.35	1.11
Sulfur	4.37	5.96	6.74	2.74
Ash	0.72	0.76	1.32	0.99
Moisture	7.60	5.69	2.24	2.86
Higher Heating Value				
MJ/kg	33.18	33.34	32.53	32.43
Btu/lb	14298	14364	14017	13972
HGI	54	39	35	55

[1]Note: Some values in table depend upon crude oil properties.
Sources: [4, 5].

With the performance characteristics, the subtle differences between types of petroleum cokes begin to emerge. The volatility differences are significant—despite the low volatility of all petroleum cokes represented in the table. The sulfur and nitrogen differences are a function of crude oil source as well as coking technology. Lower sulfur petroleum cokes, however, are more frequently used in non-fuel applications; petroleum cokes used for fuel purposes are typically high in sulfur content.

The volatility measures lead to significant considerations regarding petroleum coke structure and reactivity. Traditional methods of fuel analysis provide significant insights as shown above; more detailed analyses yield additional measures of significance.

Table 2.3. Typical Empirical Performance Parameters of Petroleum Coke

Analysis	Fuel Type			
	Delayed coke	Shot coke	Fluid coke	Flexicok
Volatility Measures				
Fixed Carbon/Volatiles Ratio	17.86	29.18	18.52	75.92
Volatiles/Fixed Carbon Ratio	0.056	0.034	0.054	0.013
H/C Atomic Ratio	0.53	0.46	0.30	0.04
O/C Atomic Ratio	0.0004	0.0086	0.0073	0.00
Pollutant Measure (kg/GJ)				
Sulfur (as SO_2)	2.63	3.57	4.14	1.69
Nitrogen	0.76	0.48	0.72	0.34
Ash	0.22	0.23	0.40	0.31
Pollutant Measure (lb/10^6 Btu)				
Sulfur (as SO_2)	6.11	8.30	9.62	3.92
Nitrogen	1.78	1.11	1.68	0.79
Ash	0.50	0.53	0.94	0.71

Basis: Table 2.2

2.2.2. *Reactivity Measures for Petroleum Coke*

Reactivity of petroleum coke, like all solid fuels, is a function of chemical structure. Recognizing that the vast majority of all petroleum coke is produced in delayed cokers, analysis focuses upon delayed petroleum coke. Reactivity measures used here include maximum volatile yield and both devolatilization and char oxidation kinetics. Black Thunder Powder River Basin (PRB) subbituminous coal and Pittsburgh #8 bituminous coal are shown, for comparison, as reference fuels.

Maximum volatile yield was measured by The Energy Institute of Pennsylvania State University [14-15] for a suite of fuels including an high volatile shot petroleum coke with 14.6 percent volatile matter (see Figure 2.1). Using a drop tube furnace at temperatures up to 1700°C (3092°F) and using air-dried fuel ground to 85 - 74 my m (140 x 200 mesh), maximum volatile yields were measured as shown below. The

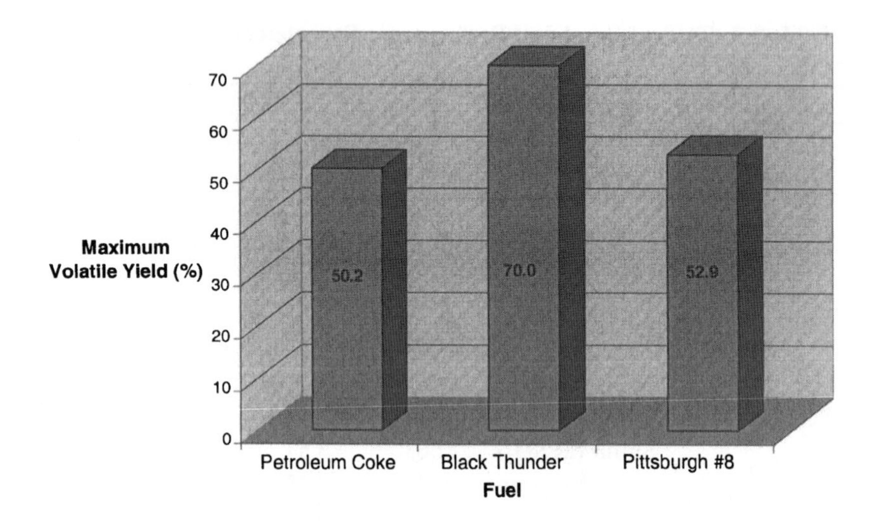

Figure 2.1. Maximum Volatile Yields for Petroleum Coke and Reference Coals
Source: [14]

relatively volatile petroleum coke exhibited a maximum volatile yield of 47.7 percent compared to 70 percent for the Black Thunder and 52.9 percent for the Pittsburgh #8 coal. Volatiles evolved from the petroleum coke at these high temperatures—which significantly exceed the temperatures employed in the proximate analysis test—included virtually all hydrogen, oxygen, and nitrogen and a some of the carbon as well.

The research by Pennsylvania State University yielded an equation for relating volatile matter in proximate analysis to maximum volatile yields as shown below:

$$MVY = 0.645*(VM_{pa}) + 39.9 \qquad [2\text{-}1]$$

Where MVY is maximum volatile yield (percent), and VM_{pa} is volatile matter measured in a standard proximate analysis. The r^2 for this equation is 0.945, over a suite of 10 fuels ranging from fresh sawdust to petroleum coke and including lignite, PRB, western bituminous, Illinois basin, and Pittsburgh seam coals. Using this equation the maximum volatile yields for the petroleum cokes shown in Table 2-2 would be 42.8 percent for delayed coke, 41.9 percent for shot coke, and 40.7 percent for flexicoke.

The volatility determination was part of a program to determine devolatilization kinetics for petroleum coke. Samples of the petroleum coke were reacted in the drop tube furnace at temperatures from 1000°C to 1700°C, in an argon atmosphere, to determine the rate of devolatilization. The procedure for calculating the kinetics was as follows:

The fuel was weighed from the feeder before and after the test, to determine the total amount of fuel feed to the drop tube furnace. Likewise, the char was carefully collected and weighed from three sources: the char deposited on top of the filter paper, the char trapped in the filter paper, and the char deposited on the probe wall. The sum of these three yields the total char collected, for a given test time. From the weight of the fuel fed (w_f) and the weight of the char collected (w_c), the percent weight loss (V) was calculated using equation 2-2:

$$V = \left[\frac{w_f - w_c}{w_f}\right] \times 100 \qquad [2\text{-}2]$$

The reactivity R, at a given drop tube furnace temperature, was calculated by equation 2-3:

$$R = \frac{\dfrac{V}{V_\infty}}{t_r} \qquad [2\text{-}3]$$

where V_∞ is the maximum percent weight loss that occurs at any drop tube furnace temperature, and t_r is the residence time of the fuel particle in the drop tube furnace. The particle residence time was calculated by dividing the length (L) that the particle travels through the region of the drop tube furnace that is at the desired temperature, by the particle velocity. The particle velocity is the sum of two components: the gas stream velocity (V_g) and the terminal velocity (V_t).

$$t_r = \frac{L}{V_g + V_t} \qquad [2\text{-}4]$$

The centerline gas velocity was approximated by doubling the bulk gas velocity. The bulk gas velocity was calculated as the volumetric gas flow rate divided by the cross sectional area of the tube inside the drop tube furnace. Stokes' law was used to calculate the terminal velocity. At a temperature of 1700 °C (3092°F), the calculated residence time for the coal particles was 186ms. The resulting devolatilization kinetics for petroleum coke are shown in Figure 2.2. The devolatilization reactivity equations among various petroleum cokes will be reasonably consistent; rather, the significant variability will come from the maximum volatile yield.

The pyrolysis/devolatilization kinetics determined for the relatively volatile petroleum coke can be compared to those for Black Thunder subbituminous coal, and Pittsburgh #8 bituminous coal. This comparison is shown in Table 2.4. Note the higher pre-exponential constant, A, and activation energy, E, associated with the petroleum coke, relative to the reference coals. The kinetic parameters shown above are consistent with the lower maximum volatility of petroleum coke.

The parameters shown in Table 2.4 can be used to evaluate the performance of fuels in specific boilers, using computational fluid dynamics (CFD) modeling, recognizing that the temperatures used to calculate the kinetics (1000°C – 1700°C) were bulk reactor temperatures rather than particle temperatures.

Char oxidation kinetics were determined at The Energy Institute of Pennsylvania State University using Thermogravimetric Analysis (TGA). Chars generated at 1700°C (3092°F) were subjected to TGA analysis in an air atmosphere, at atmospheric pressure, at temperatures ranging from 350 to 425°C (660 – 800°F). The sample was held at the

Table 2.4. Short Table of Devolatilization Kinetic Parameters of Petroleum Coke, Compared to Representative Eastern Bituminous and Powder River Basin Coals

Fuel	Pre-exponential constant, A, (1/sec)	Activation energy, E, (kJ/mol)
Petroleum Coke	104	48.19
Black Thunder	59.1	39.93
Pittsburgh #8	89.5	44.83

Source: [14]

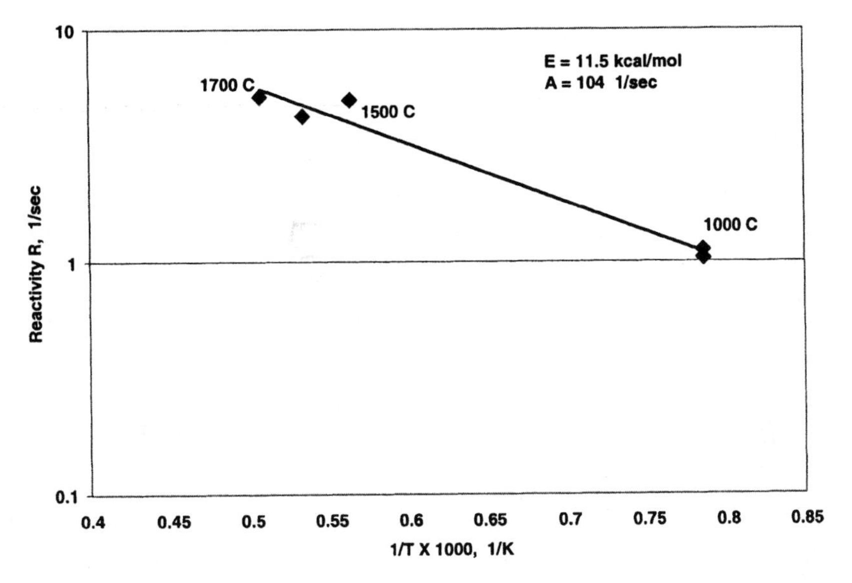

Figure 2.2. Petroleum coke pyrolysis/devolatilization kinetics
Source: [14].

desired temperature approximately 12 hours, depending on the reaction rate. Weight change data was recorded as a function of time. From this data, instantaneous reactivity (R_u) was calculated by equation 2-5:

$$R_u = \frac{1}{W_u} \cdot \frac{dW}{dt} \qquad [2\text{-}5]$$

where W_u is the dry ash free weight of unreacted char at time t; and dW/dt is the slope of the burn-off curve at corresponding time t. Reaction rates were determined from this data using a method developed by Tsai and Scaroni [16]. They showed that the reactivity of several chars was a function of burn-off and the value of reactivity can vary by a factor of two depending on the extent of burn-off at which the reactivity was calculated. Therefore, a more meaningful value than the instantaneous reactivity at a given weight loss for predicting char burn-out rates would be a reactivity value averaged over the entire char burn-out range. An average reactivity can then be calculated by equation 2-6:

$$R_{a,\,x\%} = \frac{\sum (R_u \cdot \Delta W)}{\sum \Delta W} \qquad [2\text{-}6]$$

where: R_u is the instantaneous reaction rate with respect to unreacted char, and ΔW = the weight loss occurring within each time interval. The summation of the ΔW values is equal to x, which can be anywhere from 0 to 100% depending on the range of char burn-off. For typical coal chars, Tsai and Scaroni [16] found that, for burn-offs greater than 50%, the R_a values were found to be relatively constant. Therefore, $R_{a,50\%}$ can be substituted for $R_{a,100\%}$. Figure 2.3 displays the resulting kinetics calculated for petroleum coke char.

Again the char oxidation kinetics for the petroleum coke can be compared to the char oxidation kinetics for the PRB and bituminous coals, as is shown in Table 2.5. Note, again the lower reactivity associated with the petroleum coke char relative to the subbituminous and bituminous coal chars.

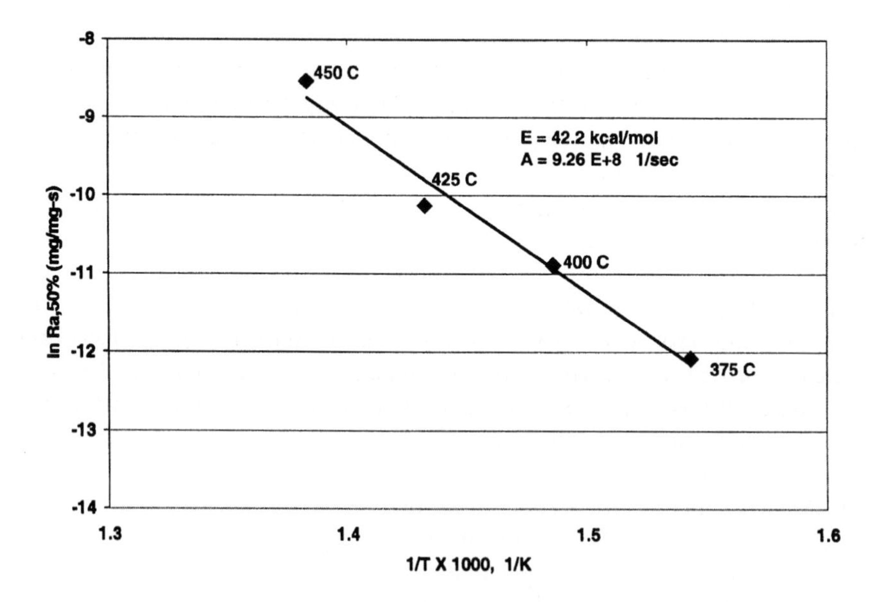

Figure 2.3. Petroleum coke char oxidation kinetics
Source: [15]

Table 2.5. Short Table of Char Oxidation Kinetic Parameters For Petroleum Coke Compared to Representative Eastern Bituminous and Powder River Basin Coals

Fuel	Pre-exponential constant, A, (1/sec)	Activation energy, E, (kJ/mol)
Petroleum Coke	9.26×10^8	176.8
Black Thunder	7.61×10^4	114.8
Pittsburgh #8	3.72×10^5	135.3

Source: [14]

The relatively low maximum volatile content and the kinetic parameters reflect the structural characteristics of the highly condensed aromatic petroleum coke. Table 2.6 presents structural data developed by The Energy Institute of Pennsylvania State University for the relatively volatile petroleum coke tested, along with the reference coals. Note that 81 percent of the carbon atoms in the petroleum coke are in aromatic structures, and that there are typically 21 aromatic carbons per cluster. This indicates that there are 5 – 6 aromatic rings per cluster compared to 3 – 4 aromatic rings/cluster in the bituminous coal and ~2 rings/cluster for the PRB subbituminous coal. It also indicates—by inference—relatively fewer functional groups and bridges available to form volatile matter. Statistical relationships between ^{13}C NMR data and maximum volatile yield suggest that the typical petroleum cokes shown in Table 2.2 will have an even higher aromaticity, and potentially a higher number of aromatic carbons/cluster than the petroleum coke tested.

The reactivity data presented in this section document the high autoignition temperature for petroleum coke, along with its relatively slow burning characteristics. The dominant combustion mechanism is the heterogeneous char oxidation mechanism. These factors become significant in the utilization methods that have been successful for this opportunity fuel.

2.2.3. Nitrogen Reactivity in Petroleum Coke

NO_x control through combustion manipulation, as well as post-combustion devices such as selective catalytic reduction systems, is of

Table 2.6. Structural Comparison of Petroleum Coke and Reference Coals Using ^{13}C NMR Analysis (distributions shown as % of total carbon atoms)

Chemical Shift	Assignment	Petroleum Coke	Pittsburgh #8 Coal	PRB Coal, Black Thunder
	SSB	2.97	3.99	0.57
177.5	Carboxylic acid	0	0	3.98
150.4	Phenol	0	4.53	11.44
140.0	Catechol	0	3.38	1.46
125.7	Aromatic	74.05	59.34	44.54
52.0	Methoxy	0	0	4.21
40.6	C, CH	8.15	11.24	5.63
30.4	CH2	7.76	5.12	15.10
19.9	CH3	3.49	10.30	12.08
	SSB	3.58	2.09	0.99
Aromaticity	(%)	81	70	57
AC/Cl(*)		21	15	10

Note: AC/Cl is the average number of aromatic carbon atoms per cluster

Source: [15, 17]

critical importance in the use of any opportunity fuel. Opportunity fuels can improve, or impede, the ability of any combustion system to reduce NO_x emissions. Traditionally combustion manipulation involves staged air and staged fuel systems that are designed into low NO_x burners, separated overfire air (SOFA) systems, and related technologies. These technologies depend upon the release of nitrogen volatiles in a fuel-rich environment, and the subsequent conversion of nitrogen volatiles to N_2. Control of NO_x from the nitrogen in char, while beginning to be evaluated [18], is far more problematic than control of NO_x from volatile nitrogen.

Critical to NO_x control, then, is the nitrogen concentration in the fuel (see Table 2.3), the proportion of nitrogen that exists as volatile matter, and the pattern of nitrogen release from the fuel. Baxter et. al. [19] has documented that volatile nitrogen release from coals varies modestly as a function of the coal, however the Baxter data indicate that the nitrogen volatile evolution largely lags behind the release of volatile carbon and hydrogen—and release of the total mass of volatiles from the fuel mass.

Experiments conducted at The Energy Institute of Pennsylvania State University [14, 20] for a range of coals and opportunity fuels have been previously reported [15, 21]. These data show that the petroleum coke contains a higher percentage of char-nitrogen than coals or biomass fuels. The comparison between petroleum coke and the reference coals is shown in Figure 2.4. Note that the maximum volatile nitrogen yield from petroleum coke is <64 percent, compared to 75 percent for the Pittsburgh #8 coal and 80 percent for the Black Thunder PRB coal.
The computation of maximum volatile nitrogen yield was based upon the ultimate analysis of the chars generated at all drop tube furnace temperatures from 1000°C to 1700°C. Petroleum coke can have significant concentrations of fuel nitrogen measured in g/GJ (lb/10^6 Btu) as shown in Table 2.3. Further, this nitrogen can be concentrated more heavily in the char than the nitrogen in coal.

Petroleum coke volatile evolution patterns are shown as a function of temperature in Figure 2.5, presenting total volatile matter evolution along with carbon and nitrogen volatile evolution.

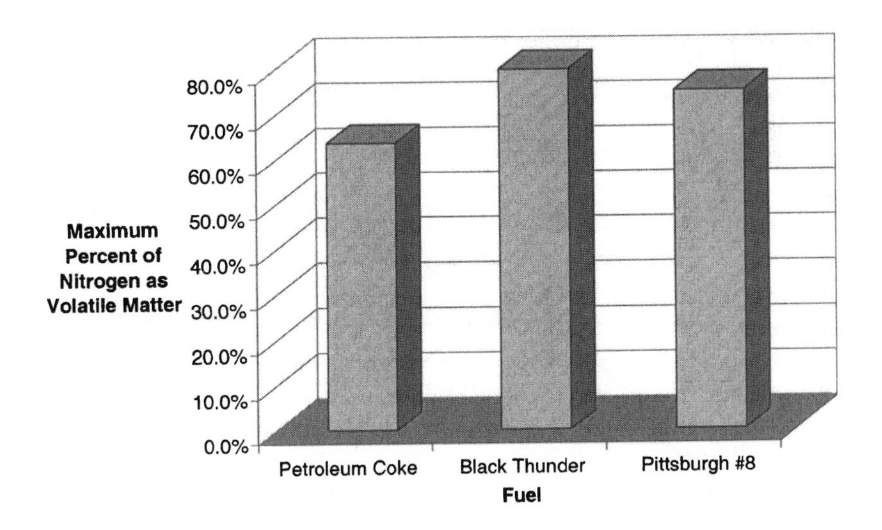

Figure 2.4. Comparison of Maximum Volatile Nitrogen Yields for Petroleum Coke, Black Thunder PRB Coal, and Pittsburgh #8 Coal
Source: [15]

Note that there is a modest lag for petroleum coke volatile nitrogen evolution in the early stages of devolatilization, and that nitrogen volatile evolution exceeds carbon volatile evolution and total volatile matter evolution at the latter stages of pyrolysis.

The patterns of nitrogen volatile evolution for petroleum coke can be compared to the reference fuels, using the data developed by The Energy Institute of Pennsylvania State University, as is shown in Figure 2.6. In this figure, the data are presented as a function of temperature. Further, for each temperature, the nitrogen/carbon (N/C) atomic ratio in the char produced by the drop tube furnace runs has been computed from ultimate analyses measured for each char sample. Where the N/C ratio increases, the data indicate that the nitrogen volatile evolution lags behind the carbon volatile evolution; the carbon volatile evolution is a reasonable proxy for total volatile matter evolution as is shown in Figure 2.5. Conversely, where the N/C ratio decreases, the nitrogen is volatilizing more rapidly than the carbon in the fuel.

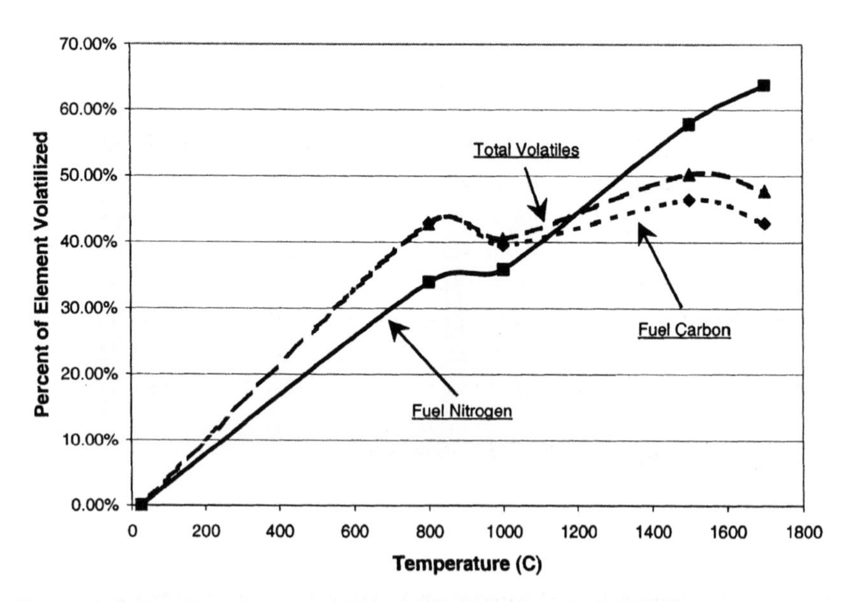

Figure 2.5. Fuel Nitrogen, Fuel Carbon, and Total Volatile Matter Evolution from Petroleum Coke as a Function of Temperature
Data developed from [14]

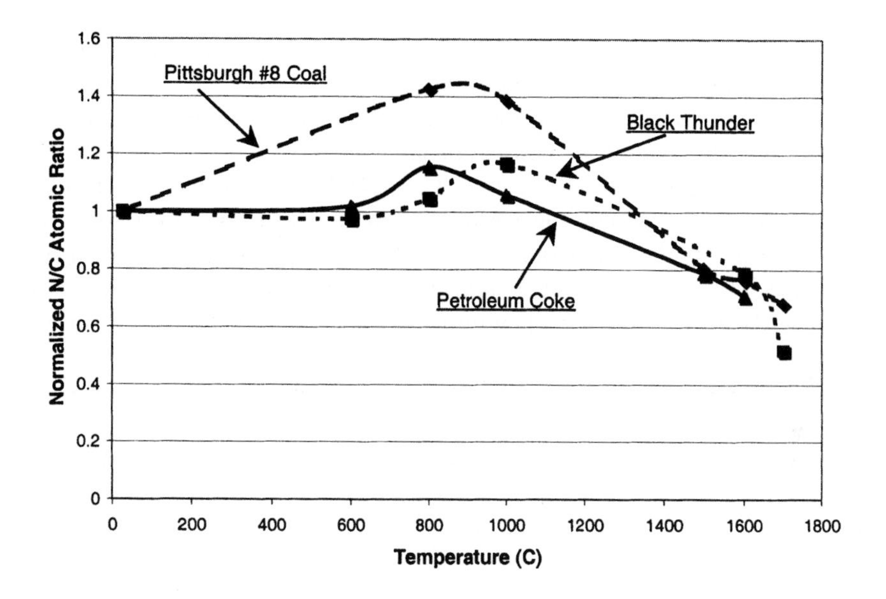

Figure 2.6. Nitrogen/Carbon (N/C) Atomic Ratios for the Char Produced in Drop Tube Furnace Experiments as a Function of Temperature
Data developed from [14].

The data in Figure 2.6 demonstrate that the petroleum coke shows more of a tendency to an initial lag in volatile nitrogen evolution than the Black Thunder subbituminous coal. However the petroleum coke shows less of a tendency for an initial lag in volatile nitrogen evolution than the Pittsburgh #8 bituminous coal. Based upon these data, and the previous volatility data, it is readily apparent that NO_x emissions from petroleum coke firing are more likely to be problematic in pulverized coal boilers with low residence times than in other types of firing systems.

2.2.4. Ash Characteristics of Petroleum Coke

Petroleum coke is, inherently, a low ash fuel; however the ash chemistry of petroleum coke remains significant due to issues of slagging and fouling, and trace metal emissions. Bryers [4, 5] has reported general ash characteristics for petroleum cokes as a function of coking method.

These data are shown in Table 2.7. Note the high concentrations of vanadium and nickel in all but the fluid coke.

As a practical matter, the concentration of vanadium, nickel, and other inorganic matter in petroleum coke is largely a function of the source of the crude oil as is shown in Table 2.8. "Sweet" (low sulfur) crude oils not only contain less ash, but can contain lower concentrations of vanadium, nickel, and other deleterious metals (see data from [4]). The vanadium content in the Venezuelan crude oil shown in Table 2.8 is particularly high, while the vanadium content in the Canadian crude oil is

Table 2.7. Typical Ash Characteristics of Petroleum Coke

Analysis	Fuel Type			
	Delayed coke	Shot coke	Fluid coke	Flexicoke
Elemental Composition (wt %)				
SiO_2	10.1	13.8	23.6	1.6
Al_2O_3	6.9	5.9	9.4	0.5
TiO_2	0.2	0.3	0.4	0.1
Fe_2O_3	5.3	4.5	31.6	2.4
CaO	2.2	3.6	8.9	2.4
MgO	0.3	0.6	0.4	0.2
Na_2O	1.8	0.4	0.1	0.3
K_2O	0.3	0.3	1.2	0.3
SO_3	0.8	1.6	2.0	3.0
NiO	12.0	10.2	2.9	11.4
V_2O_5	58.2	57.0	19.7	74.5
Ash Fusion Temperatures (°C)				
Reducing				
Initial Def.	1599	1436	1378	1538
Spherical	1599	1599	1386	1538
Hemispherical	1599	1599	1439	1538
Fluid	1599	1599	1474	1538
Oxidizing				
Initial Def.	1374	1259	1095	749
Spherical	1425	1429	1155	790
Hemispherical	1431	1471	1183	853
Fluid	1433	1471	1224	1311

Source: [4]

quite low. As will be shown in subsequent sections of this chapter, mercury concentrations in petroleum coke are consistently and significantly lower than those associated with crude oils. This results from the thermal processes used in oil refining, and the consequent volatilization of the mercury in the crude oil. By the time petroleum coke is formed, the mercury has been removed by the refining processes.

Because of the high concentrations of vanadium and nickel in petroleum coke from most regions, more detail is useful. Both vanadium and nickel are typically found in porphyrin complexes [5]. Other elements such as calcium, iron, and magnesium are found in both organic and inorganic forms [5]. Aluminum, silicon, and sodium are typically found only in inorganic forms [5]. Vanadium presents potential problems in terms of high temperature corrosion. Nickel also presents potential problems during combustion and flyash management.

Because the nickel and vanadium are of primary concern, it is useful to note the potential nickel and vanadium compounds that can be formed during combustion of petroleum coke. Both nickel and vanadium are lithophiles. Table 2.9 identifies such compounds along with their melting and decomposition temperatures.

Table 2.8. Trace Metal Concentrations in Crude Oil From Various Locations (mg/kg oil)

Metal	Source of Crude Oil			
	Typical	Libya	Venezuela	Alberta
As	0.0006 – 1.1	0.077	0.284	0.0024
Co	<1.25	0.032	0.178	0.0027
Cr	0.002 0.02	0.0023	0.430	---
Cu	N/A	0.19	0.21	---
Hg	0.09	---	0.027	0.084
Mn	0.4	0.79	0.21	0.048
Ni	14 – 68	49.1	117	---
Sb	0.002 – 0.8	0.055	0.303	---
Se	0.03 – 1	1.10	0.369	0.0094
V	15 – 590	8.2	1,100	0.682
Zn	N/A	62.9	0.692	0.670

Sources: [22 - 26]

Table 2.9. Melting and Decomposition Temperatures of Nickel and Vanadium Compounds Potentially Formed from Petroleum Coke Combustion

Compound	Symbol	Melting and Decomposition Temperature (°C)
Nickel Oxide	NiO	2090
Nickel Sulfate	$NiSO_4$	Decomp 840; → NiO
Vanadium Trioxide	V_2O_3	1970
Vanadium Tetraoxide	V_2O_4	1970
Vanadium Pentoxide	V_2O_5	675 – 690
Sodium Metavanadate	$Na_2O{\cdot}V_2O_5$	630
Sodium Pyrovanadate	$2Na_2O{\cdot}V_2O_5$	640
Sodium Orthovanadate	$3Na_2O{\cdot}V_2O_5$	850
Sodium Vanadyl Vanadates	$5Na_2O{\cdot}V_2O_4{\cdot}5V_2O_5$	625
Sodium Vanadyl Vanadates	$5Na_2O{\cdot}V_2O_4{\cdot}11V_2O_5$	535
Nickel Sulfides	NiS	1832
	Ni_2S_4	800
	Ni_2S	645
Nickel Pyrovanadate	$2NiO{\cdot}V_2O_5$	900
Nickel Orthovanadate	$3NiO{\cdot}V_2O_5$	900
Ferric Metavanadate	$Fe_2O_3{\cdot}V_2O_5$	860
Ferric Vanadate	$Fe_3O_4{\cdot}2V_2O_5$	855

Source: [4]

Note the wide array of potential compounds that can be formed depending upon temperature and combustion conditions.

The vanadium trioxide (V_2O_3) and vanadium tetraoxide (V_2O_4) compounds are typically associated with approximate unburned carbon levels in the flyash ≥10 percent. This is consistent with the porphyrin structure of the vanadium in the crude oil, and in the petroleum coke. Nickel can occur as a sulfide, an oxide, or in a vanadate (e.g., nickel pyrovanadate) depending upon combustion condition. For example, reducing conditions promote formation of sulfide compounds. Some of the compounds shown (e.g., the sodium vanadate compounds) are more commonly associated with oil firing than with petroleum coke firing, as a consequence of the low ash content in petroleum coke and the low sodium content in the petroleum coke ash. Further, most of the vanadium in products of petroleum coke combustion remains in a lower oxidation state (e.g., V_2O_3 and V_2O_5).

2.2.5. Conclusions Regarding Petroleum Coke as a Fuel

The various types of petroleum coke, then, exhibits significant potential as opportunity fuels. Economically, they are a low cost energy source and have the potential to remain so as a consequence of increasing coking capacity worldwide. While there are differences between petroleum coke fuels as a function of coking method and crude oil source, all such fuels are high in calorific content and low in moisture and ash. They are highly aromatic and less reactive than coals. The combustion mechanisms that dominate petroleum coke combustion are associated with heterogeneous gas-solids reactions.

Fuel grade petroleum cokes are typically high in sulfur content, and can be high in nitrogen content. Further, the nitrogen is more typically associated with the char fraction of the fuel than the volatile content of petroleum coke. The nitrogen is somewhat more reactive than some bituminous coals, but slightly less reactive than PRB coal nitrogen.

While petroleum cokes are low in ash, they are typically high in vanadium although this depends upon the source of the crude oil. The nickel and vanadium are found in porphyrin structures and, in combustion systems, can be found in a wide variety of compounds within the flyash. These characteristics dictate the methods for petroleum coke utilization in large utility boilers, industrial boilers, cement kilns, and other combustion systems.

2.3. Petroleum Coke Utilization in Cyclone Boilers

Cyclone boilers, installed by electric utilities until about 1975 in order to utilize slagging coals typically found in the Midwestern US, are distinct combustion systems that are highly favorable to cofiring petroleum coke. Coal crushed typically to 6.35 mm x 0 mm (¼" x 0") is introduced into a barrel typically 2.4 – 3.0 m (8 – 10 ft) diameter and 3.0 m – 3.7 m (10 – 12 ft) long where combustion occurs in a highly intense environment. Normally 2 – 14 cyclone barrels are mounted horizontally for a given furnace; the largest cyclone boiler has 23 cyclone barrels.

Each cyclone barrel can support about 40 – 50 MW_e of generation. The intensity is reflected by typical heat release rates in the cyclone barrels – nominally 19 - 32 GJ/m^3-hr (500,000 – 850,000 Btu/ft^3-hr). During combustion, a slag layer forms in the cyclone barrel and flows from the barrel to the furnace and then to a slag tank below the furnace. Typically ~70 percent of the inorganic matter in the coal is removed as slag tapped from the barrels and the furnace; ~30 percent of the inorganic matter is removed as flyash.

Cyclone boilers comprise about 9 percent of the coal-fired boiler capacity in the US, and are also used in large industrial complexes such as the Eastman Kodak manufacturing park in Rochester, NY. Cyclone boilers are typically fired as baseload units. Typically they generate significant NO_x emissions, on the order of 0.52 – 0.65 kg/GJ (1.2 – 1.5 $lb/10^6$ Btu) unless NO_x controls are added such as separated overfire air (SOFA), reburn technology, or selective catalytic reduction (SCR) technology. The high NO_x emissions, coupled with developments that facilitated firing slagging coals in pulverized coal (PC) boilers, ultimately terminated further development of this technology.

Cyclone boilers are highly favorable for cofiring petroleum coke. Because cyclone boilers were developed for a crushed fuel rather than a pulverized fuel, and because cyclone boilers were designed to remove inorganic matter as slag, they have become known for fuel flexibility. At the same time, the requirement to have sufficient ash in the fuel to form a slag layer—typically >5 percent ash in the coal—limits the use of opportunity fuels in these combustion systems to cofiring applications. Further, cyclones require significant volatile matter concentrations to promote ignition and combustion within the cyclone barrel. While petroleum coke contains significant concentrations of sulfur, the coal originally used in cyclone boilers typically was (and frequently remains) a high sulfur coal. Consequently many of these boilers are equipped with scrubbers. Further most cyclone boilers are concentrated in the Midwestern United States, and have ready access to petroleum coke from inland refineries and refineries with access to the Mississippi River transportation system.

Petroleum coke has been cofired in several such units [21, 27]. The impact of petroleum coke cofiring in cyclone boilers has been detailed in two such units: Paradise Fossil Plant (PAF), owned by Tennessee Valley Authority (TVA), and Bailly Generating Station (BGS), owned by Northern Indiana Public Service Co. (NIPSCO), a NiSources Company.

2.3.1. Petroleum Coke Cofiring at Paradise Fossil Plant

Petroleum coke was cofired in Paradise Fossil Plant (see Figure 2.7) Unit #1, a 700 MW_e cyclone boiler equipped with 14 cyclone barrels, each 3.05 m (10 ft) in diameter. These cyclones are equipped with radial feeders. The boiler generates 611.6 kg/sec (4,850,000 lb/hr) of 166.6 bar/565°C/538°C (2415 psig/1050°F/1000°F) steam. The unit is equipped with a wet scrubber for SO_2 emissions control. At the time of testing, the typical fuel burned at this installation was Western Kentucky high sulfur coal, with characteristics as shown in Table 2.10. Also shown in Table 2.10 is the analysis of the petroleum coke burned during the test period.

Detailed petroleum coke cofiring tests took place at PAF in October and November 1995, preceding an extended run with this opportunity fuel. Some 12 tests were conducted including 5 baseline tests and 7 cofiring tests with petroleum coke cofired at 12 and 23 percent of the heat input to the boiler, summarized in Table 2.11.

The impact of cofiring petroleum coke on boiler capacity and efficiency was negligible, neither improving nor reducing boiler efficiency [28]. Petroleum coke did increase the flame temperature in the cyclone barrels at PAF, as is shown in Figure 2.8. This increase resulted from the calorific value of the fuel fed to the boiler, however it was moderated by the combustion rate for the petroleum coke that was slower than that for the high volatile, high sulfur Western Kentucky coal.

The use of petroleum coke had a significant beneficial impact reducing NO_x emissions as is shown in Figure 2.9. The 23 percent petroleum coke cofiring on a heat input basis caused up to a 20 percent reduction in NO_x emissions. The variability, or scatter, within the data

Figure 2.7. Paradise Fossil Plant, a large cyclone installation of TVA

Table 2.10. Characteristics of Fuels Burned at Paradise Fossil Plant Petroleum Coke Tests

Parameter	Western Kentucky Coal	Petroleum Coke
Proximate Analysis (wt %)		
Fixed Carbon	44.3	85.6
Volatile Matter	37.4	8.1
Ash	8.2	0.9
Moisture	10.1	5.4
Higher Heating Value		
MJ/kg	27.61	33.14
Btu/lb	11,860	14,240
Sulfur and Nitrogen		
Sulfur (S) (%)	3.35	5.4
Fuel S, kg/GJ (lb/10^6 Btu)	1.215 (2.82)	1.633 (3.79)
Fuel Nitrogen (N) (%)	1.39	1.4
Fuel N, kg/GJ (lb/10^6 Btu)	0.504 (1.17)	0.422 (0.98)

Sources: [28]

Table 2.11. Summary of Results from Paradise Fossil Plant Tests

Test No.	Firing Rate MW_{th}/hr (10^6 Btu/hr)	Petroleum Coke %, heat input basis (*)	Excess O_2 (%)	Boiler Thermal Efficiency (%)
1	1727 (5901)	0	3.4	87.7
2	1391 (4753)	0	3.4	88.0
3	1800 (6150)	0	3.5	87.9
4	1717 (5860)	0	3.6	89.8
5	1744 (5958)	12	3.4	87.7
6	1747 (5966)	23	3.5	87.9
7	1750 (5976)	23	3.5	87.8
8	1811 (6181)	23	3.5	87.7
9	1191 (4070)	23	4.0	88.7
10	1811 (6189)	23	3.6	88.0
11	1811 (6184)	12	3.5	88.1
12	1822 (6227)	0	3.6	87.5

Source: [28]

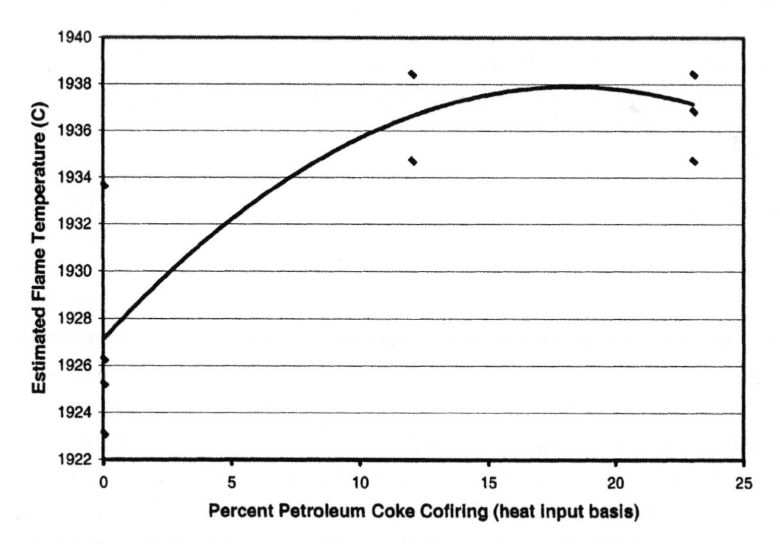

Figure 2.8. Influence of Petroleum Coke Cofiring on Estimated Flame Temperature at Paradise Fossil Plant

Source: [28]

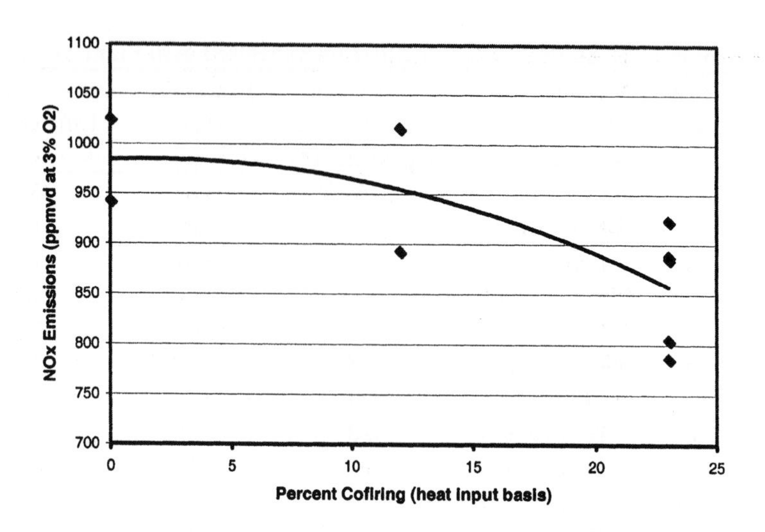

Figure 2.9. NO_x Reduction at Paradise Fossil Plant as a Function of Petroleum Coke Cofiring
Source: [28]

coupled with the relatively few combustion tests results in a lack of statistical significance. The trend, however, is readily apparent.

The firing of petroleum coke at Paradise Fossil Plant had no significant impact on SO_2 emissions. The scrubber overcame the increased sulfur content of the petroleum coke compared to the high sulfur Western Kentucky coal. No data were obtained concerning reagent consumption in the scrubber, however the coal alone had 1.215 kg/GJ (2.82 lb/10^6 Btu) sulfur while the maximum petroleum coke/coal blend had 1.299 kg/GJ (3.01 lb/10^6 Btu). The increase in sulfur concentration was not sufficient to cause excessive consumption of scrubbing reagent.

The testing demonstrated that cofiring petroleum coke with coal at Paradise Fossil Plant reduced the concentrations of mercury, chromium, lead, zinc, and cadmium in the fuel fed to the boiler [28]. The only metals exhibiting a significant increase in concentration in the fuel blend were vanadium and nickel. Because of the increase in concentration of vanadium and nickel, attention was given to the consequent partitioning and speciation of these metals found in abundance in the petroleum coke. Vanadium and nickel concentrations

were measured in the fuel, the flyash, and the slag. Flyash samples analyzed included economizer hopper ash, air heater hopper ash, flyash captured by the electrostatic precipitator, and flyash captured by the scrubber. Figure 2.10 presents increased partitioning of vanadium and nickel towards the cyclone slag as a function of petroleum coke cofiring. Note that, when cofiring petroleum coke is at least 10 percent on a mass basis, or 12 percent on a Btu basis, over 75 percent of the vanadium and over 70 percent of the nickel are concentrated in the slag.
Microprobe analysis conducted by Hazen Research of Golden, CO was used to identify the form of vanadium in the flyash and the slag. In the slag, the vanadium was found in silicate complexes and iron oxide compounds. There were no discrete vanadium pentoxide (V_2O_5) occurrences in the slag. Vanadium in the flyash was found largely in large or small silicate glassy particles. Some vanadium was also found in association with carbon, either associated with carbon particles or as inclusions within carbon compounds. It is generally recognized that vanadium in the presence of significant concentrations of unburned

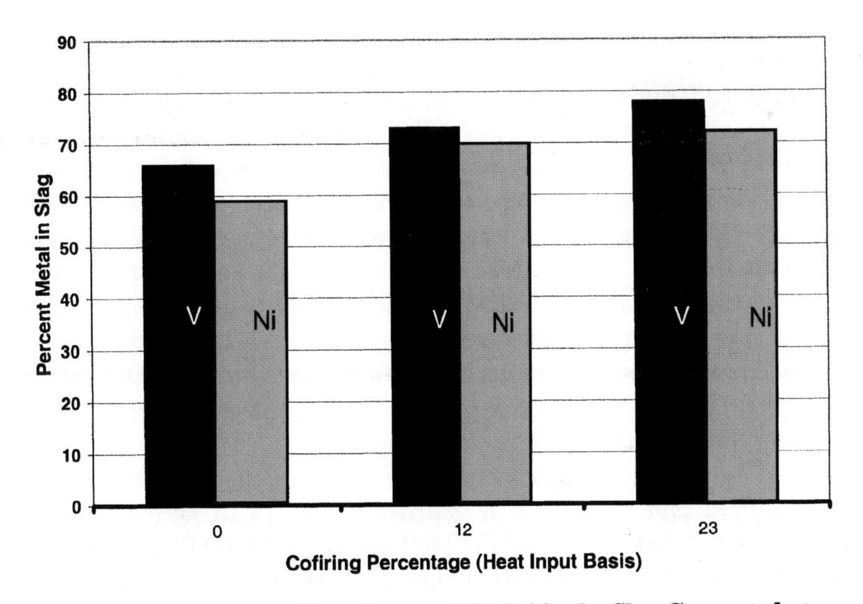

Figure 2.10. Percentage Vanadium and Nickel in the Slag Generated at Paradise Fossil Plant as a Consequence of Petroleum Coke Cofiring
Source: [28]

carbon does not oxidize completely to V_2O_5, but remains in a lower and less corrosive oxidation state (e.g., V_2O_3, V_2O_4). The nickel found in both the slag and the flyash was largely NiO [21]. These results are dependent upon the fact that the petroleum coke was cofired with a midwestern bituminous coal; cofiring with western coals would achieve different results due to the calcium based ash.

The cofiring of petroleum coke at Paradise Fossil Plant demonstrated key benefits to the use of this opportunity fuel in cyclone boilers. Testing at the Bailly Generating Station of NIPSCO further reinforced and extended these findings.

2.3.2. Cofiring Petroleum Coke at the Bailly Generating Station

Petroleum coke was cofired in more extensive tests at boiler #7 of Bailly Generating Station as part of the biomass cofiring program developed through the Cooperative Agreement between USDOE and EPRI. These tests, conducted at a NIPSCO location, were designed to demonstrate the benefits and issues associated with firing petroleum coke with coal, urban wood waste with coal, and the triburn effect of firing petroleum coke and urban wood waste with coal. A four month program was instituted during 1999, and 56 individual tests were performed. During these tests, a baseline was developed and then extensive cofiring and triburn testing was performed with petroleum coke percentages ranging from 10 percent to 25 percent on a mass basis, or 12 percent to 29 percent on a heat input basis. Urban wood waste percentages were 5, 7.5, and 10 percent on a mass basis or about 2.5, 3.5, and 5 percent on a calorific value basis [15, 29 – 31]. This concept of burning petroleum coke with another opportunity fuel—a supplier of volatile matter as well as heat input—is not unique to Bailly Generating Station. Owensboro Municipal Utilities fired petroleum coke with tire-derived fuel (TDF) at its Elmer Smith Station Unit #1 also [32].

Bailly Generating Station (BGS) Boiler #7 is a 160 MWe (net) cyclone fired unit. At MCR it generates some 151.3 kg/sec (1.2×10^6 lb/hr) of 165.5 bar/538°C/538°C (2400 psig/1000°F/1000°F) steam. At the time of testing the plant fired a blend of 70 percent high sulfur Illinois basin coal and 30 percent low sulfur western bituminous Shoshone coal. For emissions control the BGS boilers are equipped with both electrostatic precipitators and a Pure Air wet scrubber of Air Products and

Chemicals for SO_2 control. Consequently the plant is well equipped to fire high sulfur fossil fuels. During the testing, petroleum coke was fired directly with the blended Illinois basin and Shoshone coal. During many tests, the petroleum coke was blended with urban wood waste for firing with the Illinois basin and western bituminous coals. Table 2.12 presents the salient fuel characteristics of the four fuels burned at Bailly Generating Station. Because 56 individual tests were conducted over a 3-month period, a substantial body of data were obtained concerning the impact of petroleum coke cofiring on cyclone operations.

The tests demonstrated that petroleum coke exerted a distinct benefit to boiler operations, improving efficiency and reducing the volumetric flow of fuel to the boiler. The overall regression equation describing the triburn impact on boiler efficiency is shown below:

$$\eta_{tb} = 86.75 - 0.068(\%W) + 0.051(\%PC) \qquad [2\text{-}7]$$

Table 2.12. Characteristics of Fuels Burned at Bailly Generating Station Tests

Parameter	Fuel			
Proximate Analysis (wt %)	Petroleum Coke	Illinois #6 Coal	Shoshone Coal	Urban Wood
Fixed Carbon	78.27	41.95	42.15	12.50
Volatile Matter	13.90	34.43	37.56	52.56
Ash	1.34	9.66	5.63	4.08
Moisture	6.48	13.97	14.66	30.78
Higher Heating Value				
MJ/kg	33.31	25.87	25.37	13.48
Btu/lb	14,308	11,113	10,900	5,788
Sulfur and Nitrogen				
Sulfur (%)	5.11	3.45	0.74	0.07
Sulfur, kg/GJ	1.53	1.34	0.29	0.05
Sulfur (lb/10^6 Btu)	3.55	3.10	0.68	0.12
Fuel Nitrogen (%)	1.23	1.22	1.44	1.00
Fuel Nitrogen, kg/GJ	0.37	0.47	0.57	0.75
Fuel Nitrogen (lb/10^6 Btu)	0.86	1.10	1.32	1.73

Source: [15]

where η_{tb} is thermal efficiency of the boiler during triburn testing (including petroleum coke and urban wood waste cofiring testing), %W is percent wood in the blend on a mass basis, and %PC is percent petroleum coke in the fuel blend on a mass basis. The coefficient of determination—r^2—for this equation is 0.86. The probabilities of random occurrence of the results are as follows: overall equation, 2.28×10^{-20}; %W, 2.73×10^{-9}; and %PC, 5.42×10^{-17}.

The technical and economic consequence of the improved boiler efficiency—up to 1.25 percent when cofiring 25 percent petroleum coke on a mass basis or 29 percent petroleum coke on a heat input basis—included reduced fuel flow to the bunkers and boiler. This, in turn, reduced the load on conveyors, fans, and other related mechanical systems. The improved efficiency resulted in a slightly reduced house load as a function of firing petroleum coke with coal.

These efficiency gains associated with cofiring petroleum coke occurred despite increases in unburned carbon (UBC) in the flyash. When cofiring petroleum coke with coal, the unburned carbon in the flyash increased from an average of 9.1 percent (baseline – 0 percent petroleum coke) to an average of 13.5 percent UBC with 10 percent petroleum coke cofiring (mass basis) and an average of 22.6 percent UBC when cofiring 20 percent petroleum coke (mass basis) [30]. However these high UBC levels are somewhat misleading. With petroleum coke having <2 percent ash, and with 70 percent of the inorganic material leaving the boiler as slag, there is less dilution of UBC with inorganic material in the flyash, and there is less flyash. The overall impact of flyash UBC on efficiency was negligible.

Because BGS employs a wet scrubber, SO_2 emissions were not evaluated. Focus was given to NO_x emissions, CO and hydrocarbon (THC) emissions, SO_3 emissions, and trace metal emissions. NO_x emissions were of foremost importance. The following regression equation was derived to describe the impact of petroleum coke on NO_x emissions:

$$NO_x\ (kg/GJ) = 0.317 - 0.0046(\%W) - 0.0045(\%PetC) + 0.0005(Lm) + 0.0117(EO_2) \quad [2\text{-}8]$$

NO_x (lb/10^6 Btu) = 0.691 – 0.0101(%W) – 0.0098(%PetC) + 0.0005(L) + 0.0255(EO_2) [2-9]

Where %W is percent urban wood waste on a mass basis, %PetC is percent petroleum coke on a mass basis, L is load, expressed in tonne/hr for equation 2-8, and 10^3 lb/hr of main steam flow for equation 2-9, and EO_2 is excess oxygen measured on a total basis and reported in the control room. With 56 degrees of freedom, these equations have an r^2 of 0.700. Some 70 percent of the NO_x emissions can be explained with this equation. The probability that the equation, in either SI or English units, occurred randomly is 5.4×10^{-13}. The probability that the urban wood waste impact is random is 0.0014 and the probability that the petroleum coke impact is random is 1.5×10^{-11}. Clearly the petroleum coke helped reduce NO_x emissions.

While 70 percent of the NO_x emissions can be explained by this equation, 30 percent remain unexplained. Included in the 30 percent are the synergistic effects between the wood waste and the petroleum coke. When firing a blend of 22.5 percent petroleum coke/7.5 percent wood waste, NO_x emissions were reduced by 30 percent—to 0.409 kg/GJ (0.95 lb/10^6 Btu). This exceeds the values predicted by equation [2-8]. The reality of this synergy is shown by equation [2-10], where the petroleum coke and urban wood waste are combined as an opportunity fuel on a 2:1 and 3:1 mass ratio basis.

NO_x (kg/GJ) = 0.6167 + 7×10^{-5}(OF%^{2}) – 0.0071(OF%) [2-10]

Where OF% is the mass percentage of opportunity fuel. There is no difference between the 2:1 and 3:1 opportunity fuel blend, however both petroleum coke and urban wood waste must be present. The r^2 for this equation is 0.85, and the probability that this equation is created as a random occurrence is 9.64×10^{-11}.

CO, THC, and SO_3 emissions showed only minor influences as a consequence of cofiring petroleum coke as is shown in Figure 2.11. Note that petroleum coke slightly increased CO emissions and slightly decreased THC emissions. Only the combination of petroleum coke and wood waste increased SO_3 emissions above noise levels, and the total SO_3 emissions under these conditions were <14 ppmvd. Petroleum coke also

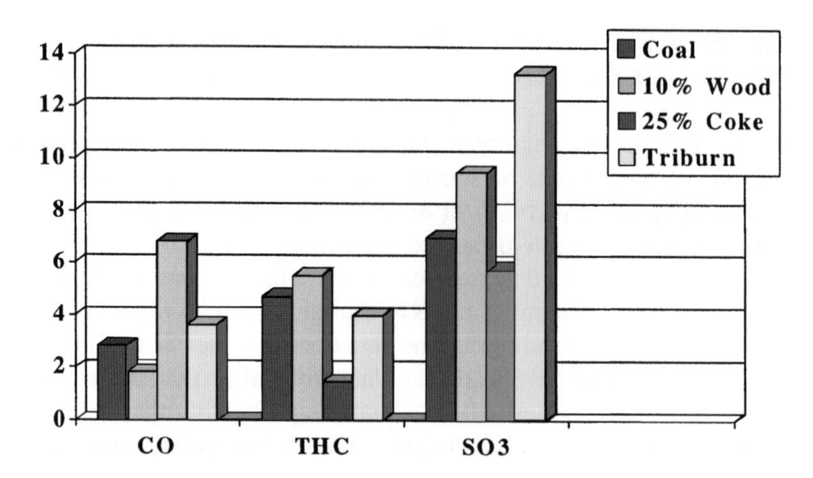

Figure 2.11. CO, Hydrocarbon, and SO_3 Emissions at the Bailly Generating Station Cofiring Tests (values in ppmvd at 3% O_2)
Source: [15]

reduced the trace metals found in the feed to BGS #7. Trace metal reductions occurred for arsenic, cadmium, chromium, lead, and mercury.

Vanadium and nickel were the only metals increasing in concentration as a consequence of cofiring petroleum coke, as expected. Table 2.13 documents the concentrations of trace metals in the petroleum coke fired at BGS, relative to the coal and urban wood waste fired at that

Table 2.13. Concentrations of Trace Metals in Fuels Burned at Bailly Generating Station Tests (values in mg/kg fuel)

Metal	Fuel			
	Petroleum coke	Low sulfur coal	High sulfur coal	Urban wood waste
Arsenic	0.337	1.402	2.194	2.145
Chromium	4.676	8.250	20.361	6.570
Lead	2.182	4.267	3.154	2.922
Mercury	0.016	0.020	0.033	0.013
Nickel	134.04	7.396	12.33	2.645
Vanadium	326.37	11.95	17.21	3.06

Source: [29]

location. The concentrations of trace metals shown in Table 2.13 indicate that, on a heat input basis, the petroleum coke further reduces the trace metal concentrations in the fuel fed.

The impact of petroleum coke, and urban wood waste, on the trace metals in the total fuel blend—calculated in mg/kg or kg/GJ ($lb/10^6$ Btu)—is shown for mercury. Mercury concentrations in the blended fuels can be calculated by equations below [30].

Hg (mg/kg) = 0.292 – 1.3x10-4(%PetC) – 1.6x10-4(%W) [2-11]

$$Hg(kg/GJ) = [2.64x10^{-6} – 1.9x10^{-8}(\%PetC) – 2.9x10^{-9}(\%W)]x0.43 \quad [2\text{-}12]$$

$$Hg(lb/10^6\ Btu) = 2.64x10^{-6} – 1.9x10^{-8}(\%PetC) – 2.9x10^{-9}(\%W) \quad [2\text{-}13]$$

The fates of vanadium and nickel were also examined during the BGS test program. Metal balances were conducted for the tests when cofiring petroleum coke with coal—with and without the addition of urban wood waste. On average, and with little deviation, the vanadium partitioned equally between flyash and slag; only half of the vanadium ended up in the flyash. Only 25 percent of the nickel ended up in the flyash with 75 percent reporting to slag [15, 29-31].

Using a flyash sample with 0.26 percent vanadium, Foster Wheeler analyzed the forms of vanadium using X-ray diffraction. The Xray diffraction system examined the sample for 18 hours to produce usable patterns. Despite the low concentrations of vanadium in the flyash relative to the typical requirements of X-ray diffraction analysis, very discernable patterns were found, providing at least a qualitative assessment of the flyash. The analysis indicated significant concentrations of V_2O_3 and V_2O_4 as well as V_2O_5 in the flyash sample. There was more evidence of vanadium in the lower oxidation states than in the pentoxide form. Electron Spectroscopy for Chemical Analysis (ESCA) experiments performed on the flyash at the University of Utah. ESCA sees only the surface of particles, rather than the total concentrations within particles as is found in X-ray diffraction. The ESCA experiments found the vanadium in the form of V_2O_5, indicating that the surface of the particles is where the vanadium pentoxide is concentrated.

The cofiring of petroleum coke with coal, and the trifiring of petroleum coke with urban wood waste and coal at Bailly Generating Station boiler #7 clearly documented the benefits of using this opportunity fuel in cyclone boilers. Boiler efficiency improved, leading to modest reductions in house loads associated with fuel handling equipment and with forced draft and induced draft fans. NO_x emissions decreased as a function of firing petroleum coke and—most interestingly—as a function of firing petroleum coke and urban wood waste with the coal. Carbon monoxide, hydrocarbon, and sulfur trioxide emissions were largely unaffected by petroleum coke or urban wood waste. Trace metal emissions were decreased. The Bailly tests documented the clear advantage of petroleum coke in cyclone boilers.

2.4. Cofiring Petroleum Coke in Pulverized Coal Boilers

Pulverized coal (PC) boilers are the most common large combustion systems for the generation of electricity in the US and the industrialized economies of the world. PC boilers include both wall-fired boilers and tangentially-fired (T-fired) boilers; wall-fired boilers include both front wall and opposed wall configurations. Like cyclone boilers, they have the potential to use petroleum coke as a fuel provided that a sulfur dioxide scrubber has been installed on the unit. The vast majority of the petroleum coke fired in the USA is burned in PC boilers due to their dominance of the industry. Again the low volatility in the petroleum coke limits its use in PC boilers; the typical cofiring percentage is on the order of 20 – 30 percent (calorific value basis).

2.4.1. General Opportunities and Constraints with PC Boilers

Cofiring petroleum coke with coal in PC boilers has the economic potential associated with reducing fuel costs identified previously. However, there are certain limitations with PC boilers. Many of these units sell flyash as pozzolanic material under the American Society for Testing and Materials (ASTM) Specification C-618. ASTM C-618 limits the unburned carbon content in the flyash [32]. About 25 percent of the flyash generated in the US is sold under ASTM Specification C-618. The ability to sell flyash—or the inability to sell flyash—can impact plant economics by >$1 - $3 million annually. Few cyclone boiler power plants sell flyash.

Low NO_x firing has been significantly advanced for PC boilers capitalizing upon high volatile release at the base of the flame—and capitalizing upon staged air-staged fuel combustion. While these concepts have been developed for cyclone firing, they have been sufficiently advanced for PC firing that several tangentially-fired units using PRB coals have achieved NO_x emissions <64.5 g/GJ (0.15 lb/10^6 Btu) [33].

The constraints on PC cofiring of petroleum coke are more significant than in the case of cyclone firing. For all pulverized coal units, the Hardgrove Grindability Index (HGI) becomes critical, because pulverizing petroleum coke to an appropriate size (e.g., 80 percent <74 my m or 200 mesh) may be uneconomical for certain petroleum cokes with low HGI values. Flame stability for wall-fired PC boilers becomes a critical issue, and can limit the percentage of petroleum coke in the total fuel mix. Consequently the results firing petroleum coke in PC boilers is mixed. Three case studies—Widows Creek Fossil Plant (WCF) of TVA, Limestone Unit #2 of Reliant Energy, and W.A. Parish Station of Houston Lighting and Power highlight some of the potentials and limitations of petroleum coke utilization in large PC utility boilers.

2.4.2. *Cofiring Petroleum Coke at Widows Creek Fossil Plant*

Widows Creek Fossil Plant consists of two plants: A plant with 6 smaller wall-fired boilers, and B plant with two large T-fired boilers. Boiler #7 has a net generating capacity of 518 MW_e and boiler #8 has a net generating capacity of 531 MW_e. Typically they are operated at about 400 MW_e (net). These units generate 154.4 bar/538°C/538°C (2400 psig/1000°F/1000°F) steam. Both units are equipped with a SO_2 scrubber, sufficient to handle the increased emissions resulting from petroleum coke cofiring [34]. Testing of cofiring petroleum coke was conducted on document the impact on boiler efficiency for Units #7 and #8 respectively. both boilers at 20 and 25 percent (mass basis, or 24 and 29 percent, calorific value basis), with a brief test at 35 percent in unit #8. Characteristics of the petroleum coke and coal fired at the WCF cofiring tests are shown in Table 2.14.

Table 2.14. Fuel Characteristics for the Widows Creek Petroleum Coke Tests

Parameter	Petroleum Coke	Coal
Proximate Analysis (wt %)		
Volatile Matter	9.25	23.31
Fixed Carbon	84.66	56.81
Moisture	5.56	9.51
Ash	0.54	10.67
Ultimate Analysis (wt %)		
Carbon	84.4	66.4
Hydrogen	3.6	4.5
Oxygen	0.14	4.6
Nitrogen	0.97	1.25
Sulfur	4.8	3.07
Moisture	5.56	9.51
Ash	0.54	10.67
Higher Heating Value		
MJ/kg (Btu/lb)	32.84 (14150)	26.94 (11067)

Source: [34]

The impact of the petroleum coke on efficiency was immediate, and significant. Equations 2-14 and 2-15

$$\eta_{\#7} = 99.5 + 0.029(\%PetC) - 0.453(EO_2\%) - 0.060(T_{ah}) \quad [2\text{-}14]$$

and

$$\eta_{\#8} = 98.9 + 0.029(\%PetC) - 0.377(EO_2\%) - 0.060(T_{ah}) \quad [2\text{-}15]$$

Where %PetC is percent petroleum coke on a mass basis, $EO_2\%$ is percent excess O_2 at the air heater exit on a dry basis, and T_{ah} is the temperature of the air exiting the air heater (°C). Conversion of this equation from SI to English units requires changing the T_{ah} term from 0.06 to 0.03. The r^2 for these equations exceeds 0.99 [34].

The use of petroleum coke improved boiler efficiency; at 20 percent petroleum coke, boiler efficiency improved by over 0.5 percent and net station heat rate decreased by >70kJ/kWh (65 Btu/kWh). This efficiency improvement could be attributed to reduced losses associated with hydrogen and moisture in the fuel. The efficiency improvements came despite increases in unburned carbon in the flyash. Table 2.15 summarizes the impacts of petroleum coke cofiring on unburned carbon in the flyash for both units. Given that flyash is 80 percent of the solid products of combustion from a PC boiler, these increases were significant. For plants selling flyash under ASTM C-618, these would be unacceptable results. For plants without flyash sales, these results can be accepted given the efficiency improvements.

Testing showed no appreciable impact on flame temperature, upper furnace temperature, or flame intensity when firing petroleum coke with high sulfur coal. Flame intensity decreased when firing petroleum coke briefly with low sulfur coal. Similarly, there were no operability issues; main steam and reheat steam temperatures did not decrease, and maintaining load was not at issue [34].

Environmental consequences of cofiring petroleum coke were also measured. SO_2 emissions were a function of the scrubber performance, not the fuel blend. SO_3 emissions increased modestly as a function of cofiring petroleum coke, although statistical analysis of the impact of petroleum coke on SO_3 emissions is inconclusive. Since SO_3 affected opacity, a portion of the WCF tests were performed with injection of hydrated lime as a means for SO3 control. Hydrated lime injection proved successful [34].

NO_x emissions increased modestly during the cofiring of petroleum coke at WCF, as is shown in Table 2.16. Given the high concentrations of char-bound nitrogen in petroleum coke and the rates of volatile nitrogen release as shown previously, this result is not unexpected.

Table 2.15. Average Unburned Carbon in the Flyash at Widows Creek

Boiler	Percent Petroleum Coke (mass basis)	Average % Unburned Carbon in Flyash
7	0	NA
7	20	6.33
7	25	9.29
8	0	1.39
8	20	9.17

Source: [34]

Table 2.16. Average NO_x Emissions During Full Load Petroleum Coke Cofiring Testing at Widows Creek Fossil Plant

Boiler	Percent Petroleum Coke	NO_x Emissions (kg/GJ)	NO_x Emissions (lb/10^6 Btu)
7	0	0.14	0.32
7	20	0.17	0.39
7	25	0.17	0.40
8	0	0.11	0.25
8	20	0.12	0.28
8	35	0.13	0.31

Source: [34]

The NO_x emissions associated with PC firing are fundamentally different from those associated with cyclone firing. The baseline NO_x emission level is much lower, and the impact of staging depends more heavily upon fuel volatility due to the presence of the flame within the primary furnace.

Vanadium oxidizing to V_2O_5 was not a problem at the WCF tests. The high concentrations of unburned carbon in the flyash, particularly at the higher petroleum coke cofiring percentages, inhibited formation of V_2O_5 during these tests [34].

2.4.3. Petroleum Coke Cofiring at Limestone Unit #2

Cofiring petroleum coke with lignite and PRB coal was tested at Limestone Unit #2, an 820 MW_e boiler located 125 miles north of Houston, TX. Limestone Unit #2 is a T-fired boiler with a eight corner furnace configuration. Like the WCF boilers, Limestone Unit #2 is a subcritical boiler, generating 154.4 bar/538°C/538°C (2400 psig/1000°F/1000°F) steam. The unit was retrofitted with the Foster Wheeler Tangential Low NOx (TLN3) system in the year 2000. Subsequently it was tested with Gulf Coast Lignite, Caballo Rojo PRB coal, and blends of these coals with 20 - 23 percent shot petroleum coke (calorific value basis) [33].

Petroleum coke cofiring with lignite was not successful. The petroleum coke did not blend well with the lignite in the pulverizers due to differences in hardness; significant quantities of petroleum coke were rejected in the mills. Shot coke has a Hardgrove Grindability Index that can be as low as 35. Further, the cofiring of 23 percent petroleum coke with lignite reduced main steam temperatures slightly as a function of a cleaner furnace. No data have been published on the impacts of petroleum coke on boiler efficiency.

Average full load NO_x emissions increased from 82 g/GJ (0.193 lb/10^6 Btu) when firing lignite alone, to 108 g/GJ (0.25 lb/10^6 Btu) when adding 20 percent petroleum coke (mass basis) to the fuel feed. This was later reduced to 90 g/GJ (0.21 lb/10^6 Btu) for the cofiring case [33].

Petroleum coke cofiring with PRB coal also was not successful. Average main steam temperatures decreased from 539°C (1003°F) when firing 100 percent PRB coal, to 512°C (954°F) when cofiring petroleum coke at a 20 percent level. Reheat temperatures decreased from 543°C (1009°F) when firing PRB coal, to 520°C (968°F) when cofiring petroleum coke at the 20 percent level. NOx emissions were 58.9 g/GJ (0.137 lb/10^6 Btu) when firing 100 percent PRB coal, and 71.8 g/GJ (0.167 lb/10^6 Btu) when firing the petroleum coke/PRB blend [33]. Like the Widows Creek tests, NO_x emissions increased as a function of cofiring petroleum coke with coal in a PC boiler.

2.4.4. *Cofiring Petroleum Coke at W.A. Parish Station*

W.A. Parish Generating Station boiler #8 is a 555 MWe (net) tangentially-fired boiler supplied by Combustion Engineering (now Alstom) and commissioned in 1982. It is fired with Powder River Basin coal, and equipped with roller type pulverizers from CE [35]. The unit has the capacity to generate 530 kg/s (4.2x10^6 lb/hr) of 165 bar/540°C/540°C (2400 psig/1000°F/1000°F) steam [36]. In 1995 tests were conducted at this generating station in collaboration with EPRI.

The testing at W.A. Parish focused upon mill performance, operational issues, and emissions resulting from cofiring 10 – 15 percent petroleum coke (heat input basis) with PRB coal. The differences in the fuel were dramatic. The heating value of the petroleum coke was 33.29 MJ/kg (14,300 Btu/lb) compared to 19.56 MJ/kg (8,400 Btu/lb) for the PRB coal. The moisture content of the petroleum coke was <5 percent, while the moisture content of the PRB coal was about 30 percent. The sulfur concentration was 4 – 5 percent for the petroleum coke, and 0.3 –

0.5 percent for the PRB coal. Consequently the sulfur concentrations were 3.0 kg/GJ (6.99 lb/10^6 Btu) as SO_2 for the petroleum coke, and 0.51 kg/GJ (1.19 lb/10^6 Btu) for the PRB coal. The HGI was 52+ for the petroleum coke and 63 for the PRB coal [36].

Operationally the test was successful despite difficulties in maintaining a consistent fuel blend. Despite the differences in HGI, the differences in moisture content, and the potential for dusting, mill performance during the cofiring tests was essentially similar to that of the baseline testing, except for a single mill where deterioration occurred [36]. Furnace operations were essentially not impacted by the blending. Steam attemperation was similarly largely not impacted by cofiring petroleum coke, although reheat attemperation was slightly higher. No unusual slagging occurred during the test period [36].

Loss on ignition (LOI) or unburned carbon in the flyash was significantly impacted by cofiring. LOI at W.A. Parish Unit #8 is typically <1 percent. During the cofiring tests LOI averaged 5 percent, and periodically reached 15 percent [36].

Emissions were impacted somewhat by cofiring petroleum coke. SO_2 concentrations at the inlet to the scrubber were elevated, as would be expected, from cofiring petroleum coke with low sulfur PRB coal. The scrubber was able to handle the increased SO_2 at the inlet. The scrubber was also tested using dibasic acid (DBA) as an additive to enhance its performance. That test was quite successful. NO_x emissions increased by about 5 percent when cofiring petroleum coke at 10 – 15 percent on a heat input basis. They were well within permit limitations, however.

The results of the testing at W.A. Parish demonstrated that cofiring petroleum coke with PRB coal could be a practical approach to fuel costs, although it would require attention to fuel handling and careful consideration of efficiency and emissions issues.

2.4.5. Conclusions Regarding Cofiring Petroleum Coke in PC Boilers

The data above demonstrate that petroleum coke can have certain economic advantages in PC boilers, particularly in terms of improving boiler efficiency and reducing fuel costs. Further, in many boilers, there will be no impact on main steam or reheat steam temperature and consequently no impact on turbine efficiency. The data on operational issues demonstrate, however, that the impact of petroleum coke on PC boilers is quite site-specific. Care must be taken to evaluate these impacts on any given location. The impact of petroleum coke cofiring on flyash may or may not be significant, depending upon whether or not the flyash is sold as pozzolanic material.

The impact of petroleum coke cofiring on airborne emissions from PC boilers is not as favorable as for cyclone boilers. While there are numerous reports of units where NO_x emissions did not increase when petroleum coke was introduced with coal, there is potential for this outcome. That potential is consistent with the fuel characteristics of petroleum coke. Such potential may or may not be significant depending upon whether a boiler has an associated selective catalytic reduction (SCR) system for NO_x control. Sulfur emissions also can increase, although this can be managed by scrubber technology. Nevertheless the many PC boilers cofiring petroleum coke with coal speak to its advantages as an opportunity fuel. However care must be paid to its use.

2.5. Petroleum Coke Utilization in Fluidized Bed Boilers

Circulating and bubbling fluidized bed boilers have proven readily adaptable to the combustion of many opportunity fuels including petroleum coke. The emergence of this technology in the past 25 years has given users of petroleum coke and other opportunity fuels a significant combustion tool to increase energy source flexibility.

In fluidized bed technology, the long residence times and high rates of heat transfer to the fuel particles makes them suitable for low volatile solids. At the same time their use of limestone in the bed, which subsequently calcines to calcium oxide or free lime (CaO), provides a mechanism for combustion of high sulfur fuels. The free lime reacts with SO_2 formed during the combustion process to produce calcium sulfate ($CaSO_4$). In the past decade, circulating fluidized bed (CFB) combustion has rapidly dominated this field. At the same time low combustion temperatures of about 815°C – 870°C (1500°F – 1600°F) not only optimize the calcining and sulfation reactions, but also work with staged combustion processes in the CFB systems to minimize NO_x formation.

2.5.1. Introduction

Petroleum coke can be fired in combination with coals or heavy oils in CFB boilers; alternatively it can be fired as the sole fuel. Examples of petroleum coke firing in fluidized bed boilers are shown in Table 2.17.

Table 2.17. Representative Circulating Fluidized Bed Boilers Firing Petroleum Coke or Petroleum Coke/Coal Blends

Owner	Steam Capacity (kg/sec)	Steam Capacity (10^3 lb/hr)	Main Steam Conditions (bar/°C)	Main Steam Conditions (psig/°F)
Purdue University	25.2	200	45/440	650/825
City of Manitowoc (100% pet coke)	25.2	200	67/485	975/905
Fort Howard Paper #3 (85% pet coke)	40.4	320	103/510	1500/950
Fort Howard Paper #5 (100% pet coke)	40.4	320	103/510	1500/950
Hyundai Oil, Korea (100% pet coke)	30.3	240	108/520	1565/970
University of N. Iowa (50 – 70% pet coke)	13.2	104.5	52/332	750/630
NISCO (2 reheat boilers, 100% pet coke)	104.0	825	112/540/540	1625/1005/1005
JEA, Jacksonville, FL (2x300 MW_e boilers firing coal and/or pet coke)	251.4	1993.6	172/538/538	2500/1000/1000

Sources: [38-40].

Other examples of CFB boilers firing petroleum coke alone or in combination with other fuels include installations at the Nova Scotia Power Pt. Aconi Generating Station, at Gulf Oil (now Chevron) in California, Oriental Chemical Industries in Korea, General Motors in Michigan, the Petrox refinery in Chile, a paper mill in Kattua, Finland, and numerous other sites. Most of these units fire a blend of petroleum coke and coal, with petroleum coke being the dominant fossil fuel.

The principle benefits associated with combusting petroleum coke in fluidized bed boilers include high boiler efficiencies and availabilities, and control of airborne emissions. The most significant issues associated with fluidized bed combustion of petroleum coke involve ash chemistry—management of limestone addition for optimized sulfur capture with minimum limestone cost—and management of vanadium-limestone interactions that cause agglomeration and fouling of heat transfer surfaces [41-42]. Conn [42] has identified many of the solid products of CFB combustion, as shown in Table 2.18. The low melting points of some of these compounds including sodium sulfate, vanadium pentoxide, and calcium and sodium metavanadate, indicate some of the agglomeration issues that can exist.

The advantages and issues of firing petroleum coke in fluidized bed boilers is best illustrated by a case study—the two reheat boilers installed by NISCO at Lake Charles, LA. These are among the early reheat CFB boilers. A second installation at the Northside Generating

Table 2.18. Possible ash constituents formed during CFB combustion of petroleum coke

Compound	Symbol	Melting Point °C	Melting Point
Calcium sulfate	$CaSO_4$	1450	2642
Nickel oxide	NiO	2090	3794
Sodium sulfate	Na_2SO_4	880	1616
Vanadium trioxide	V_2O_3	1970	3578
Vanadium tetraoxide	V_2O_4	1970	3578
Vanadium pentoxide	V_2O_5	680	1274
Calcium metavanadate	$CaO \cdot V_2O_5$	780	1432
Sodium metavanadate	$2Na_2O \cdot V_2O_5$	630	1166
Nickel pyrovanadate	$2NiO \cdot V_2O_5$	900	1650
Ferric metavanadate	$Fe_2O_3 \cdot V_2O_5$	860	1580

Source: [42]

Station of Jacksonville Electric Authority also provides insights into the use of petroleum coke in CFB boilers.

2.5.2 Firing Petroleum Coke in the NISCO CFB Boilers

The Nelson Industrial Steam Company (NISCO) is a partnership formed between Citgo Petroleum Company, Conoco Inc., Vista Chemical Company, and Gulf States Utilities—now part of Entergy. This partnership, formed in 1988, embarked upon construction of a cogeneration facility to supply process steam to Vista Chemical Company as well as power to the partners and the grid. This project was a repowering effort, using existing turbine-generators.

The facility installed by Foster Wheeler included two CFB boilers, each generating 103.95 kg/sec (825,000 lb/hr) of 112.1 bar/540°C (1625 psig/1005°F) main steam and 91.7 kg/sec (727,300 lb/hr) of 32 bar/540°C (464 psig/1005°F) reheat steam under base conditions. When operated producing extraction steam, the unit was designed to produce 81.6 kg/sec (647,300 lb/hr) of 28.4 bar/540°C (412 psig/1005°F) of reheat steam [43].

The NISCO project is designed to fire <6.35 mm (<¼ in) petroleum coke fired through the front wall of the lower furnace region of the boiler. Crushed limestone is fed through ports adjacent to the petroleum coke penetrations. Preheated primary air is fed through the air distributor, creating a dense fluidized bed at the bottom of the furnace. Secondary air is introduced above the fuel feed points. The products of combustion exiting the top of the furnace are ducted to three steam-cooled cyclones that also serve as primary superheaters. Solids are recovered in the cyclones and dropped to the J-valves. Solids leave the J-valves and are introduced into the INTREX™ heat exchangers functioning as secondary superheaters. INTREX™ heat exchangers are non-fired fluidized beds that capitalize upon the high rates of heat transfer between solid particles and the superheater tubes. Solid particles leave the INTREX™ heat exchangers and are returned to the furnace in the typical recirculation patterns associated with CFB combustion.

Solid products of combustion are removed either as spent bed material or as flyash. Spent bed removal is accomplished by two stripper/coolers that selectively remove coarse spend bed material from the furnace in order to control solids inventory in the CFB, remove oversized particles that could otherwise defluidize a portion of the furnace cross section, and cool the coarse spent bed material and—in the process—recover heat from the reject bed material. Flyash is captured by a reverse air baghouse supplied by Environmental Elements [43].

The NISCO project burns delayed coke from the local refineries; the typical fuel characteristics of this fuel are shown in Table 2.19. It utilizes locally available limestone, and performs final limestone preparation on-site.

Overall the project has met or exceeded its performance targets as shown in Table 2.20. Very low NO_x emissions were achieved.

Table 2.19. Petroleum Coke Fired at the NISCO Project

Analytical Parameter	Performance fuel	Range of Coke
Ultimate Analysis (wt %)		
Carbon	79.74	75-86
Hydrogen	3.31	3.0-3.6
Oxygen	0.00	0.0-0.1
Nitrogen	1.61	1.3-1.9
Sulfur	4.47	3.4-5.3
Moisture	10.60	5.5-15.0
Ash	0.27	0.0-0.5
Higher Heating Value (MJ/kg)	31.22	29.24-33.65
Higher Heating Value (Btu/lb)	13451	12600-14500
Vanadium (mg/kg in ash)	2000	500-2000

Sources [44-45]

Table 2.20. The Measured Performance of the NISCO Project

Parameter	Actual Results	Guarantee
MCR Main Steam Capacity, kg/s (lb/hr)	104 (825,000)	104 (825,000)
Boiler Thermal Efficiency (%)	90.3	83.7
Airborne Emissions		
NO_x, kg/GJ (lb/10^6 Btu)	<0.065 (<0.15)	0.258 (0.60)
CO, kg/GJ (lb/10^6 Btu)	<0.004 (<0.01)	0.026 (0.06)
VOC, kg/GJ (lb/10^6 Btu)	0.0004 (0.001)	0.3 (0.008)
TSP Particulate to Baghouse, kg/GJ (lb/10^6 Btu)	4.30 (10.0)	13.55 (31.5)

Source: [46]

The NISCO project boiler thermal efficiencies result largely from a low stack temperature—measured at 138°C (280°F) and the reduction of losses associated with moisture in the fuel and hydrogen in the fuel when compared to typical coals. The low stack temperature results from the ability to remove SO_2 from the flue gas. The low losses associated with moisture and hydrogen are inherent to petroleum coke firing.

The NISCO project, then, demonstrated the potentials associated with firing petroleum coke in CFB boilers. Further, the reliability of the plant measured by availability and capacity factors has been substantial. During the first four years of operation, plant availability exceeded 90 percent, and CFB boiler availability averaged 92.7 percent. Plant capacity factors averaged over 85 percent [46]. At the same time, however, ash agglomeration occurred in the J-valves, the cyclone barrel walls, the furnace heat recovery area, and in the INTREX™. This agglomeration problem appears to be characteristic to units firing 100 percent petroleum coke, and is not significant in units cofiring petroleum coke with higher ash coals [46]. Improvements have been made to the NISCO units to minimize if not eliminate these issues.

2.5.3. Petroleum Coke Firing at the Northside Generating Station of Jacksonville Electric Authority

The Northside project, a Clean Coal Technology Demonstration Project funded by both Jacksonville Electric Authority (JEA) and the U.S. Department of Energy (USDOE) involves repowering units #1 and #2 with new 300 MW_e CFB boilers. The project produces a combined 600 MWe of electricity for this municipal utility—the eighth largest public power organization in the USA.

The JEA project involves a reheat boiler design similar to that employed at the NISCO project. The Foster Wheeler CFB boilers are equipped with steam-cooled cyclones to accomplish additional heat recovery, and are also equipped with INTREX™ heat exchangers to maximize heat recovery from the solids [40, 47-48]. The repowering project is designed for firing both coal and petroleum coke. The Clean Coal Technology project design criteria includes NO_x emissions of 0.039 kg/GJ (0.09 lb/10^6 Btu), 0.065 kg/GJ (0.15 lb/10^6 Btu) SO_2, and 0.005 kg/GJ (0.011 lb/10^6 Btu) particulate matter. To accomplish these objectives, ammonia injection has been added to the design, with the injection point being between the primary furnace and the cyclone. Further, a polishing scrubber has been added for SO_2 control, permitting the use of 8 percent sulfur petroleum coke without excessive consumption of limestone in the fluidized bed [47]. The polishing scrubber—a spray dryer/baghouse combination—raises the sulfur capture level from the typical 90 – 95 percent for CFB combustion to 98 percent, when firing 8 percent sulfur petroleum coke [40].

Construction began on this project in 1999, and the units have since been brought on line. Environmental performance has been most impressive. These units again demonstrate the potential for firing petroleum coke in the generation of electricity through fluidized bed combustion.

2.6. Other Opportunity Fuel Uses of Petroleum Coke

Petroleum coke has also been fired in the Polk County, FL Clean Coal Technology Demonstration—in a 250 MW_e (net) integrated gasification-combined cycle (IGCC). Typical blends are about 55 percent petroleum coke-45 percent coal. This installation employs a Texaco oxygen-blown gasifier to generate a synthesis gas that is then burned in a combustion turbine. The project has been quite successful in demonstrating the flexibility of petroleum coke in power generation. However the Polk County project only highlights other applications for petroleum coke in the energy arena. The Polk County project is one of several gasification projects worldwide that employs petroleum coke as a part of the total feedstock. Similarly petroleum coke has been fired successfully at the Wabash River Clean Coal IGCC project.

Petroleum coke has been studied extensively, and periodically employed, as an additive for the manufacture of metallurgical coke for the steel industry [49-50]. This practice, ongoing for several decades, typically mixes 5 – 40 percent (weight basis) petroleum coke with coal in the coking ovens. The addition of petroleum coke reduces the reactivity of the metallurgical coke to CO_2 and improves its mechanical strength—making these cokes superior in performance to cokes made with coal alone [49]. Modest amounts of petroleum coke (e.g., ~3 percent) are sufficient to produce the desired improvements [49].

Petroleum coke has also been proposed in petroleum coke-water slurries [51-52] and petroleum coke-oil slurries [53]. Both of these slurries are considered to be replacements for heavy oil fired in existing boilers and furnaces. The coke-water slurries are more appropriate for cofiring than for single fuel firing due to the lack of volatile matter in the fuel. The coke-oil slurries are proposed as a means for extending oil supplies and firing a high calorific value in boilers. The low ash content of the petroleum coke makes this option attractive, and the oil in the slurry provides sufficient volatile matter to support ignition and combustion.

Petroleum coke, then, is a highly flexible and useful opportunity fuel that can be used not only in conventional cyclone, PC, and fluidized bed boilers; but it can also be used in gasification systems, as a feedstock for metallurgical coke for the steel industry, and as a base fuel in coke-water slurries or coke-oil slurries. With increased production of this refinery byproduct resulting from increased demand for gasoline, increased availability and use will follow.

2.7. Opportunity Fuels Related to Petroleum Coke

Heavy oils, and pitches from refining, all exhibit characteristics similar to petroleum coke. These characteristics result from the parent material—crude oil. Orimulsion™ also exhibits many of the characteristics of petroleum coke. Heavy oils, pitches, and Orimulsion™ all serve as opportunity fuels to greater or lesser extents. Pitches and heavy oils have been shown to be successful feedstocks for fluidized bed combustion [37,54]. While these oil-derived products have high volatile matter contents, they exhibit the high vanadium and nickel concentrations typically associated with petroleum coke and other crude-oil derived products. Of particular interest is the potential for using the pitch resulting from hydrotreating the Canadian tar sands [54]. An analysis of tar sands pitch is shown in Table 2.21. Note the comparison between this analysis and the petroleum coke analyses shown in Tables 2.2 and 2.7. The pitch is higher in volatile matter, and calorific value. Further, it is significantly lower in sulfur content than most petroleum cokes. The ash is extremely high in iron content (Fe_2O_3), and also contains significant concentrations of vanadium and nickel [54]. The base/acid ratio approaches 2.0. This fuel fires as a substitute solid fuel.

Table 2.21. Properties of Canadian Tar Sands Pitch from Hydrotreating

Parameter		Value
Proximate Analysis (wt %)		
	Volatile matter	67.7
	Fixed carbon	30.2
	Ash	2.0
	Moisture	0.1
Ultimate Analysis (wt %, dry basis)		
	Carbon	86.2
	Hydrogen	7.1
	Oxygen	1.0
	Nitrogen	1.1
	Sulfur	2.8
	Ash	1.8
Calorific Value, MJ/kg (Btu/lb) as received		42.6 (18,350)
Ash Characteristics (wt % of ash)		
	SiO_2	21.4
	Al_2O_3	8.2
	TiO_2	0.7
	Fe_2O_3	44.9
	CaO	6.8
	MgO	5.4
	Na_2O	1.1
	K_2O	0.3
	P_2O_5	0.0
	NiO	1.1
	V_2O_5	3.6

Source: [54]

The tar sands pitch has been fired in a CFB pilot plant, demonstrating that it has significant potential as an opportunity fuel [54]. CFB firing of pitch from oil refining, in combination with coal, has been commercially proven at the Somedith/COF installation in Marseilles, France. That boiler cofires pitch with coal at a 50/50 ratio (heat input basis) and generates 18.1 kg/s (143,600 lb/hr) of 49 bar/430°C (710 psig/806°F) steam. It has been operational since 1993 [55].

One of the most interesting opportunity fuels related to petroleum coke is Orimulsion®, a highly volatile product produced from Venezuelan crude oil. Orimulsion® is a fuel used as a substitute for heavy oil in large utility and industrial boilers. It is comprised of 70 percent bitumen from the Orinoco belt in Venezuela, and 30 percent water. A surfactant, or stabilizing agent, is used as well. Initially the product Orimulsion® 100 was marketed; this product has been enhanced and a new formulation, Orimulsion® 400 is being produced. The new formulation involves a different surfactant and the exclusion of magnesium nitrate from the product [56-58]. Orimulsion, like petroleum coke and the heavy oils and pitches previously discussed, is derived from crude oil and related products.

Orimulsion® is now being used selectively on a worldwide basis, as is shown in Table 2.22. This use includes both large scale boilers and cement kilns where it has replaced heavy oil.

Consumption of Orimulsion® now exceeds 180 PJ/yr ($171x10^{12}$ Btu/yr) worldwide. The similarities and differences between Orimulsion® and petroleum coke can readily be determined by comparing the characteristics of this opportunity fuel, as shown in Table 2.23, to those of

Table 2.22. Representative Current Use of Orimulsion® Worldwide

Utility/Location	Capacity (MW_e)	Thousand Tonne/yr	Startup date	Fuel displaced
N. Brunswick, Canada	315	750	1994	Coal/fuel oil
Kashima Kita, Japan	120 + steam	375	1991	Fuel oil
Kashima Kita, Japan	120 + steam	375	1999	New unit
Kansai Electric, Japan	156	200	1994	Fuel oil
Hokkaido Electric, Japan	130	120	1998	New unit
Energie2, Denmark	640	1500	1995	Coal/fuel oil
Enel Brindisi, Italy	1320	1650	1997	Coal/fuel oil
Enel Fiume Santo, Italy	640	1000	1998	Coal/fuel oil
Arawak	Cement plant	40	1997	N/A

Source: [59]

petroleum coke shown previously in Tables 2.2 and 2.7. Clearly the Orimulsion® is more volatile—as is the case for pitch—based upon the fact that it is not the product of successive heating and refining to produce gasoline and other light products from crude oil. Further, as an emulsion, it is a liquid fuel rather than a solid fuel. At the same time it does carry the high concentrations of vanadium and nickel and, for a liquid fuel, it is highly aromatic as a consequence of its bitumen base.

The comparison between Orimulsion® and petroleum coke with respect to volatility is shown by the H/C and O/C atomic ratios. The H/C atomic ratio for Orimulsion® is ~2.1, while it is 0.53 for delayed coke; similarly the O/C atomic ratio for Orimulsion® is ~0.33 compared to ~0.0004 for petroleum coke. The Orimulsion® is quite similar in

Table 2.23. Typical Properties of Orimulsion® 400

Characteristic		Value
Moisture content (wt %)		27.5 – 30.2
Average droplet size (μm)		14 – 20
Density (15°C), kgm^{-3}		1.013
Apparent viscosity (30°C, 100 s^{-1}) mPas		200 – 350
Flash point, °C		>90
Pour point, °C		3
Higher heating value, MJ/kg (Btu/lb)		30.24 (13,027)
Lower heating value, MJ/kg (Btu/lb)		27.49 (11,804)
Ultimate Analysis (wt %, dry basis)		
	Carbon	60.1
	Hydrogen	10.4
	Oxygen	26.4
	Nitrogen	0.35
	Sulfur	2.85
	Ash	0.2
Metals (mg/kg)		
	Vanadium	320
	Sodium	12
	Magnesium	6
	Nickel	75
	Iron	5

Source: [56,59]

ultimate analysis and volatility to pitch from the Canadian tar sands, although its moisture content is necessarily higher. Orimulsion® has been used as a supplement to, or replacement for, both oil and coal [59]. Tests at the Texaco Montebello, CA facility and elsewhere have shown that Orimulsion® can be used as a feed to IGCC plants based upon either the Texaco gasifier or the Shell gasification process [59-60].

The performance of Orimulsion® in commercial boilers has been well documented [58,61-62]. From an operational perspective, this opportunity fuel does not degrade the performance of boilers, particularly when compared to oil. Environmental impacts also have been favorable based upon tests conducted for the Florida Power & Light Company Sanford Plant [63] and other installations [64].

2.8. Conclusions Regarding Petroleum-based Opportunity Fuels

There are several petroleum based opportunity fuels led by petroleum coke. These fuels have made their way into large utility and industrial boilers, and into process industries such as cement manufacture and steel production. The characteristics of petroleum coke indicate a fuel that is high in calorific value and carbon content, however it is not highly reactive as indicated either by maximum volatile yield, H/C atomic ratio, O/C atomic ratio, or the kinetics of devolatilization and char oxidation. On the other hand alternative petroleum based opportunity fuels such as pitch and Orimulsion® are highly reactive as indicated by volatile/fixed carbon ratios, H/C atomic ratios, and O/C atomic ratios.

The petroleum-based opportunity fuels have made their way into the energy arena. Petroleum coke can be highly useful as a supplementary fuel in cyclone, PC, and fluidized bed boilers. It can be the exclusive fuel for CFB boilers as well. Alternatively, petroleum coke can be used in IGCC applications or as an additive in the production of metallurgical coke. Petroleum coke cofiring in conventional coal-fired boilers can improve boiler efficiency and make modest reductions in parasitic power loads of a generating station. When used in cyclone boilers, it can be effective in reducing NO_x emissions along with trace metal emissions; the NO_x reduction advantage does not exist in cofiring petroleum coke with coal in PC boilers.

Pitches and other oil refining residuals can be fired successfully in CFB boilers. They provide a useful substitute for solid fuels. Orimulsion® can be used to replace either oil or coal in boiler or kiln applications. It can also be used as the feed to a gasifier supporting an IGCC generating station. Orimulsion® has no operational disadvantages relative to coal or oil, and can be fired within environmental regulations.

The petroleum-based opportunity fuels, then, provide a highly useful set of alternatives for utilities and heavy industries seeking to reduce fuel costs without encountering undue operational difficulties. They readily fit within the typical infrastructure of electricity generating stations and industrial operations.

2.9. References

1. Roskill Consulting Group. 1999. The Economics of Petroleum coke. Roskill Reports on Metals and Minerals.
2. Swanekamp, R. 2002. Return of the supercritical boiler. *Power.* 146(4): 32-40.
3. Federal Energy Regulatory Commission. 2001. FERC Form 423, Monthly Report of Cost and Quality of Fuels for Electric Plants. Washington, D.C.
4. Bryers, R. 1994. Utilization of petroleum coke and petroleum coke/coal blends as a means of steam raising. In Coal—blending and switching of low-sulfur western coals. ASME, New York. pp. 185-206.
5. Bryers, R.W. 1995. Utilization of petroleum coke and petroleum coke/coal blends as a means of steam raising. *Fuel Processing Technology.* 44:121-141.
6. Jacobs Consultancy. 2002. Outlook for high-sulfur petroleum coke – 2002: a global market in change. Jacobs Consultancy, Inc., Houston, TX.
7. Narula, R.G. 2002. Challenges and economics of using petroleum coke for power generation. World Energy Council, London, England.
8. Energy Argus. 2001. Petroleum Coke. Energy Argus, Inc. Sep 3. Report No. 01S-001.
9. Heintz, E.A. 1996. Review: The characterization of petroleum coke. *Carbon.* 34(6): 699-709.
10. Heintz, E.A. 1995. Effect of calcinations rate on petroleum coke properties. *Carbon.* 33(6): 817-820.

11. Lapades, D.N. (ed.). 1976. Encyclopedia of Energy. McGraw-Hill. New York.
12. Silva, A.C., C. McGreavy, and M.F. Sugaya. 2000. Coke bed structure in a delayed coker. *Carbon*. 38:2061-2068.
13. Tillman, David and Patricia Hus. 2000. Blending Opportunity Fuels with Coal for Efficiency and Environmental Benefit. Proc. 25th International Technical Conference on Coal Utilization and Fuel Systems. Coal Technology Association. Clearwater, FL. March 6-9. pp. 659-670.
14. Johnson, David K., S.V. Pisupati, R.S. Wasco, and B.G. Miller. 2001. Pyrolysis and Char Oxidation Kinetics. Prepared for Foster Wheeler Development Corporation as a Subcontract to the Electric Power Research Institute Biomass Cofiring Project. The Energy Institute, Pennsylvania State University, State College, PA.
15. Tillman, D.A. 2001. Final Report: EPRI-USDOE Cooperative Agreement: Cofiring biomass with coal. Contract No. DE-FC@@-96PC96252. Electric Power Research Institute, Palo Alto, CA.
16. Tsai, C.Y. and A.W. Scaroni. 1987. Reactivity of Bituminous Coal Chars During the Initial Stage of Pulverized-Coal Combustion. *Fuel*. 66:1401.
17. Clifford, D.J. 2001. 13C and 15N Nuclear Magnetic Resonance Spectroscopy. Prepared for Foster Wheeler Development Corporation as a Subcontract to the Electric Power Research Institute Biomass Cofiring Project. The Energy Institute, Pennsylvania State University, State College, PA.
18. Molina, A., E.G. Eddings, D.W. Pershing, and A.F. Sarofim. 2000. Char nitrogen conversion: implications to emissions from coal-fired utility boilers. *Progress in Energy and Combustion Science*. 26:507-531.
19. Baxter, Larry L., Reginald E. Mitchell, Thomas H. Fletcher, and Robert H. Hurt. 1996. Nitrogen Release During Coal Combustion. *Energy & Fuels*. 10(1):188-196.
20. Johnson, D.K., D.A. Tillman, B.G. Miller, S.V. Pisupati, and D.J. Clifford. 2001. Characterizing Biomass Fuels for Cofiring Applications. Proc. 2001 Joint International Combustion Symposium: Towards Efficient Zero Emission Combustion. American Flame Research Committee. Kuai, Hawaii. Sep 9 – 12.

21. Tillman, D.A. 2002. Petroleum coke as a supplementary fuel for cyclone boilers: characteristics and test results. Proc. International Joint Power Generation Conference. Phoenix, AZ. June 24 – 28. Paper 2002-26157.
22. Edwards, L.O. et. al. 1980. Trace metals and stationary conventional combustion sources, Vol 1: Technical Report. Radian Corporation, Austin, TX. For USEPA, Contract No. 68-02-2608.
23. Tillman, D.A. 1994. Trace Metals in Combustion Systems. Academic Press, San Diego.
24. Thompson, W.E. and J.W. Harrison. 1978. Survey of Projects Concerning Conventional Combustion Environmental Assessments. USEPA Report 600/7-78-139. Research Triangle Institute, Research Triangle Park, NC.
25. Golightly, D.W. 1973. Atomic Emission Spectroscopy. In Survey of Various Approaches to the Chemical Analysis of Environmentally Important Materials. (B. Geifer and J. Taylor, eds). National Bureau of Standards. Washington, D.C.
26. Ondov, J.M., W.H. Zoller, I. Olmez, et. al. 1975. Elemental Concentrations in the National Bureau of Standards' Environmental Coal and Fly Ash Standard Reference Materials. Anal. Chem., 47(7): 1102-1109.
27. Letheby, K. 2002. Utility Perspectives on Opportunity Fuels. Proc. 27th International Technical Conference on Coal Utilization and Fuel Systems. Clearwater, FL. March 4-7.
28. Tillman, D.A. 1996. Petroleum coke cofiring assessment for the Paradise Fossil Plant. Prepared for the Tennessee Valley Authority, Chattanooga, TN. Foster Wheeler Environmental Corporation, Sacramento, CA.
29. Hus, P.J. and D.A. Tillman. 2000. Cofiring multiple opportunity fuels with coal at Bailly Generating Station. *Biomass and Bioenergy*. 19(6): 385-394.
30. Tillman, D.A. 1999. Biomass Cofiring: Field Test Results. Electric Power Research Institute, Palo Alto, CA. Report TR-113903.
31. Tillman, D.A. 2002. Petroleum coke as a supplementary fuel for cyclone boilers. Proc. 27th International Technical Conference on Coal Utilization & Fuel Systems. Coal Technology Association. Clearwater, FL. March 4-7. pp. 489-502.

32. Hower, J.C., J.D. Robertson, and J.M. Roberts. 2001. Petrology and minor element chemistry of combustion by-products from the co-combustion of coal, tire-derived fuel, and petroleum coke at a western Kentucky cyclone-fired unit. *Fuel Processing Technology.* 74:125-142.
33. Pearce, R. and J. Grusha. 2001. Tangential Low NOx System at Reliant Energy's Limestone Unit #2 Cuts Lignite, PRB, and Pet Coke NOx. Proc. EPRI-DOE-EPA Combined Power Plant/Air Pollution Control Symposium. Clearwater, FL. August 20-24.
34. Tillman, D.A. 1995. Results of testing Widows Creek Fossil Plant boilers 7 and 8 cofiring petroleum coke with coal. Prepared for the Tennessee Valley Authority, Chattanooga, TN Foster Wheeler Environmental Corporation, Sacramento, CA.
35. Bergesen, C. and J. Crass (eds). 1996. Power Plant Equipment Directory. Utility Data Institute, McGraw Hill Companies. Washington, D.C.
36. O'Connor, D.C. and A.P. Machtemes. 1997. Petroleum Coke Cofiring Experience at the W.A. Parish Station. Proc. Effects of Coal Quality on Power Plants. Electric Power Research Institute. Kansas City, MO. May 20-22.
37. Darling, S.L. and I.F. Abdulally. 1997. Experience with burning refinery by-products in circulating fluidized bed boilers. Proc. Power-Gen International '97. Dallas, TX. Dec 9-11.
38. Abdulally, I.F. and K.A. Reed. 1994. Experience update of firing waste fuels in Foster wheeler's circulating fluidized bed boilers. Proc. Power-Gen Asia '94. Hong Kong. Aug 23-25.
39. Castro, A.L. and P.K. Chelian. 1996. Application of Foster Wheeler CFB boilers to burn vacuum residuals for low cost power with low emissions. Proc. Mexico Power '96. Monterrey, Mexico. Oct 8-10.
40. Dyr, R.A., A.L. Compaan, J.L. Hebb, and S. L. Darling. 2000. The JEA Northside Repowering Project: Low Cost Power and Low Emissions with CFB Repowering. Proc. PowerGen-2000. Orlando, FL. Nov 14-16.
41. Anthony, E.J., A.P. Iribarne, J.V. Iribarne, R. Talbot, L. Jia, and D.L. Granatstein. 2001. Fouling in a 160 MWe FBC boiler firing coal and petroleum coke. *Fuel.* 80:1009-1014.
42. Conn, R.E. 1995. Laboratory techniques for evaluating ash agglomeration potential in petroleum coke fired circulating fluidized bed combustors. *Fuel Processing Technology.* 44:95-103.

43. Goidich, S.J., A.R. McGee, and K. Richardson. The NISCO Cogeneration Project 100-MWe Circulating Fluidized Bed Reheat Steam Generator. Proc. 1991 International Conference on Fluidized Bed Combustion. Montreal, Canada. April 21-24. pp. 57-64.
44. Voyles, R.W. and A.L. Dougherty. 1995. An update of operational experiences at NISCO. Proc. Power-Gen Americas '95. Anaheim, CA. Dec. 5-7.
45. Tharpe, D.W., and I. Abdulally. 1997. An Update of Operating Experience Burning Petroleum Coke in a Utility Scale CFB: The NISCO Cogeneration Project. Proc. 1997 Fluidized Bed Combustion Conference. ASME.
46. Voyles, R.W. and D. Zierold. 1994. An update of operational experiences at NISCO. Proc. Power-Gen Asia '94. Hong Kong. Aug 23-25.
47. Nielsen, P.T., J.L. Hebb, and R. Aquino. 1998. Large-scale CFB combustion demonstration project. Proc. 23rd International Technical Conference on Coal Utilization & Fuel Systems. March 9-13. Clearwater, FL. Pp. 23-34.
48. National Energy Technology Laboratory. 2001. The JEA Atmospheric Fluidized Bed Combustor Clean Coal Project: Repowering Northside Units 1 and 2. National Energy Technology Laboratory, US Department of Energy, Pittsburgh, PA.
49. Alvarez, R., J.J. Pis, M.A. Diez, C. Barriocanal, C.S. Canga, and J.A. Menendez. 1998. A semi-industrial scale study of petroleum coke as an additive in cokemaking. *Fuel Processing Technology*. 55:129-141.
50. Zubkova, V.V. 1999. The effect of coal charge density, heating velocity and petroleum coke on the structure of cokes heated to 1800°C. *Fuel*. 78:1327-1332.
51. Prasad, M., B.K. Mall, A. Mukherjee, S.K. Basu, S.K. Verma, and K.S. Narasimhan. 1998. Rheology of petroleum coke-water slurry. Proc. 23rd International Technical Conference on Coal Utilization and Fuel Systems. Coal Technology Association. Clearwater, FL. March 9-13. pp. 1109-1116.
52. Vitolo, S., R. Belli, M. Mazzanti, and G. Quattroni. 1996. Rheology of coal-water mixtures containing petroleum coke. *Fuel*. 75(3):259-261.
53. Xu, H., Z. Xiang, S. Weiyi, C. Xinyu, Y. Qiang, H. Zhenyu, Z. Junhu, L. Jianzhong, W. Xiaorong, and C. Kefa. 1998. Study on flow resistance and heat transfer for petroleum coke-residual oil slurry in

pipe. Proc. 23rd International Technical Conference on Coal Utilization and Fuel Systems. Coal Technology Association. Clearwater, FL. March 9-13. pp. 1119-1128.
54. Brereton, C.M.H., C.J. Lim, J.R. Grace, A. Luckos, and J. Zhu. 1995. Pitch and coke combustion in a circulating fluidized bed. *Fuel*. 74(10): 1415-1423.
55. Gamble, R.L., S. L. Darling, and I.F. Abdulally. 1998. Utilization of petroleum coke and heavy pitches in circulating fluidized bed boilers. Proc. Heavy Oils Conference, New York. June 6-7.
56. Maruffo, F. and W. Sarmiento. 1999. The new generation of Orimulsion®: enhancing the fuel. Proc. 24th International Technical conference on Coal Utilization and Fuel Systems. Coal Technology Association. March 8-11. Clearwater, FL. pp. 757-764.
57. Marruffo, F., and W. Sarmiento. 2000. Orimulsion® new generation: new commercial test results. Proc. 25th International Technical Conference on Coal Utilization & Fuel Systems. March 6-9. Clearwater, FL. Pp. 683-694.
58. Marruffo, F., W. Sarmiento, and A. Alcala. 2001. Orimulsion®, recent commercial results. Proc. 26th International Technical Conference on Coal Utilization & Fuel Systems. March 5-8. Clearwater, FL. Pp. 681-701.
59. Marruffo, F., W. Sarmiento, and A. Alcala. 2002. Opportunities for Orimulsion® gasification via IGCC technology. Proc. 27th International Technical Conference on Coal Utilization & Fuel Systems. March 4-7. Clearwater, FL. Pp. 333-343.
60. Quintana, M.E. and L.A. Davis. 1990. Pilot plant evaluation of Orimulsion as a feedstock for the Texaco gasification process. Texaco, Inc. Montebello Research Laboratory, Montebello, CA.
61. Kennedy, B.A. 1989. Evaluation of handling and combustion characteristics of a bitumen-in-water emulsified fuel in a commercial utility boiler. Proc. Power-Gen '89. New Orleans. Dec 5-7.
62. Navas, M., A.C. Sanchez, and C.O. Gomez. 2000. Orimulsion 400 boiler ash characteristics compared to Orimulsion 100 ash. Proc. 25th International Technical Conference on Coal Utilization & Fuel Systems. March 6-9. Clearwater, FL. Pp. 741-750.
63. Entropy Environmentalists Inc. 1991. Stationary source sampling report reference no. 8165A. Florida Power and Light Company Sanford

Plant, Sanford, Florida. April 1 through 5 and 8 through 12, 1991. Entropy Environmentalists Inc., Research Triangle Park, NC.

64. Johnson, I.C., N.G. Tavel, and F. Marruffo. 1999. Enhancing the fuel: Orimulsion-400, the next generation: environmental fate and effects. Proc. 24th International Technical Conference on Coal Utilization & Fuel Systems. March 8-11. Clearwater, FL. Pp. 773-784.

CHAPTER 3: COAL-WATER SLURRIES AND RELATED WASTE COAL OPPORTUNITY FUELS

3.1. Introduction

Waste coal products have created a vast resource for numerous opportunity fuels. In the USA, about 40 percent of the 1 billion tons of coal produced annually gets processed through various coal cleaning systems. Coal cleaning is used to remove inorganic matter, pyritic sulfur, trace minerals, and other contaminants [1-3]. In the 19th and early 20th centuries, coal cleaning in both the anthracite and bituminous coal fields involved manually separating lumps of shale and slate from the combustible organic rock.

Coal mining was begun in 1760, in the USA, near Pittsburgh, PA [4]. Anthracite mining was initiated in 1775 near Pittston, PA [5]. By 1917, anthracite production exceeded 100 x 10^6 tonnes of product, mined by over 180,000 workers [5]. Bituminous coal production continued on a much larger scale, fueling the steel mills and the emerging power industry. Resulting were piles of waste—anthracite culm and bituminous gob—piles that contained both waste material and useful fuel. In the 20th century, when the mechanical separation of impurities based upon specific gravity separations and the removal of fine coal particles which are difficult to handle and dewater became the technology of choice, the bituminous coal mining industry produced vast quantities of both gob and fines. The coal fines were stored in ponds and behind dams and impoundments as coal and silt-filled semi-liquid storage systems.

Hundreds of millions, if not billions, of tons of anthracite culm and bituminous gob were produced. Rhone [6] describes the process of producing anthracite culm. For any anthracite mine, a breaker building capable of processing 1,200 – 1,500 tonnes/day is constructed. Boys aged 12 and up worked in the breaker building separating slate particles from the coal. Other material entering the culm pile would be small coal particles and fines, along with coal particles missed by the boys. Culm piles would grow with the activity of the mine, sometimes encroaching on the miners' village. When the culm pile encroached upon the village, houses were moved to make way for it. Similar processes occurred in the bituminous coal fields. Piles of coal wastes can be found throughout all coal mining and using regions of the world.

Mechanical coal cleaning has become highly developed and is largely employed in the bituminous coal industry. Cleaning systems that were developed include washing, heavy media separation (e.g., using the Chance sand flotation process), columns, froth flotation, centrifuge systems, and other techniques [1-2]. These cleaning methods all are based upon water both as a transport medium and a cleaning medium; and it is this use of water that has led to the vast ponds and impoundments of fine coal and impurities. Currently there are 2×10^9 tonnes of waste coal in such ponds and impoundments, and the industry is adding 50×10^6 tonnes/year of waste coal fines to that resource [7-8].

The scope of the fine coal waste issue is dramatic. Today, some 5 – 10 percent of the run-of-mine coal produced and sent to preparation plants is discharged as fine coal slurry and disposed of in impoundments. Since 70 percent of the coal mined in the Eastern and Midwestern US is processed in preparation plants, this represents a significant resource [9]. There are over 2,300 waste coal impoundments and settling ponds in the USA today, of which over 1,500 have been abandoned [10].

Coal fines and impoundments are found throughout the coal-using world. Significant deposits of waste coal fines exist from Turkey to Italy to Chile, and throughout other coal producing countries. The problem of dealing with coal refuse is universal to the industrialized world.

The generation of substantial quantities of waste coal products has given rise to the development of several opportunity fuels including coal-water slurries (CWS) and direct use of waste coal. These opportunity fuels are specific to locales where coal has been mined and cleaned.

3.1.1. Types of Coal-Water Slurries

CWS technology has been under development for well over a century, with the blending of pulverized coal with both water and oil. The first patents for coal-oil mixtures (COM) were issued to Smith and Munsell in 1879, as a means for extending the oil resource and reducing the costs of liquid fuels [11]. Relatively low costs of crude oil over time, however, suppressed interest in this technology. COM interest again arose in the 1970's due to the oil price shocks, and several types of slurry fuels were developed including COM, with <10 percent water in the fuel slurry, coal-oil-water (COW) suspensions of coal in fuel oil and >10 percent water, coal-methanol fuel (CMS) where fine coal particles are slurried in methanol, and various CWS formulations [11].

In the 1980's, research emphasis was placed on developing CWS formulations with high concentrations of coal, seeking an oil substitute. These formulations focused upon developing a product that handled much like #6 heavy fuel oil. Typical solids concentrations in these CWS products were 70 – 75 percent [12]. High-density CWS fuels increased the concentration of coal fines in the product from ~40 percent (wt basis) in COM to the 65 – 70 percent associated with the CWS products [13].

In the 1990's, lower density CWS fuels were investigated and developed. Their primary applications were, and remain, cofiring with coal in large utility boilers [13 – 15]. These slurries were developed to provide coal preparation plants a means for utilizing hard to dewater fines, as a means for cleaning up waste coal fines impoundments and thereby reducing safety hazards of coal mining, and as a means for supplying utilities with a low cost fuel that could also function as a NO_x trim mechanism [13-14, 16]. These slurries contained 40 – 55 percent coal fines (wt basis) compared to the concentrations found in the petroleum substitute products [14].

3.1.2. Potential Uses of CWS Fuels

CWS fuels developed for cofiring have been developed for both pulverized coal boiler firing and for cyclone boiler firing. The slurries have somewhat different characteristics, largely as a function of the boiler type. Since cyclone boilers employ a coal crushed typically to <6.35 mm (<¼"), while pulverized coal boilers require a coal pulverized to about 70 – 80 percent <74 μm (<200 mesh), the grind for a cyclone boiler can be significantly relaxed relative to that used in a pulverized coal boiler. Table 3.1 summarizes the different types of CWS product developed for use as opportunity fuel.

CWS fuels have numerous potentially attractive features. They can be environmentally beneficial, both from the coal preparation perspective and from the combustion perspective. They can be a low cost fuel, particularly if used near the point of production. They integrate readily into existing generating stations. They also have the potential to extend petroleum supplies.

In reality, however, CWS represents a family of fuels. They are variable based upon the sources of fine coal, the size fractions of the fine coal, the extent of coal processing, the density of the slurry, and related factors. If produced from coal pond fines these fuels can be high in ash and high in certain trace metals. In order to consider these opportunity fuels, this chapter first considers the basic published properties and then examines case studies of CWS production and utilization. Following the consideration of CWS, the chapter discusses analogous fuels and then the direct use of waste coal products in electricity generating stations.

3.2. Fuel Characteristics of Coal-Water Slurries

The characteristics of CWS fuels are dependent significantly on the type of product sought, and produced. Further, the characteristics depend significantly on the initial coal mined, the method and extent of coal cleaning, and the disposal practices.

3.2.1. Characteristics of High Density CWS Fuels

High-density CWS fuels, discussed both in Steam [12], and in Morrison et. al. [15], have been produced on both a research and development basis by a joint venture between Occidental Petroleum and Combustion Engineering—OXCE—and by Carbogel, a Swedish development company. Carbogel worked with Foster Wheeler in the development of high-density CWS fuels.

Table 3.1. Typical Parameters for Coal-Water Slurry Fuels

	Fuel Application		
	Fuel Oil Substitution	Pulverized Coal Cofiring	Cyclone Cofiring
Historical Perspective			
Time frame	Pre 1990	Post 1990	Post 1990
Driving force	Reduction in fuel oil consumption	Reduced fuel cost, NO_x trim	Reduced fuel cost, NO_x trim
Combustion Parameters			
Primary fuel	CWS	Pulverized coal	Crushed coal
Secondary fuel	Fuel oil, gas	CWS	CWS
Boiler bulk gas residence time	<1 sec	1 – 3 sec	<1 sec
Atomized CWS droplet size	Critical; (d_{50}<80μm)	Less stringent than fuel oil case	Less stringent than fuel oil case
CWS injection location	Burner	Burner	Scroll section of cyclone
Ash handling	Required	Available	Available
CWS Preparation Parameters			
Solids loading	High; 55-70%	50-60%	40-60%
Additive package	Critical (dispersant, pH modifier, stabilizer	Typically not used	Typically not used
Stability requirements	Long term required	Short term (days)	Short term (days)
Apparent viscosity	<500 cp @ 100/s; pseudoplastic flow behavior	Less critical (typically 200 cp @ 100/s)	Less critical (typically 200 cp @ 100/s)
Coal particle size distribution	Broad; d_{50}<20μm	Typically d_{50} ~45μm	60% <250 μm (60 mesh) with d_{50} 297 μm (50 mesh)
Coal ash	<5 wt%, high ash fusion temp	<15 wt %	6 – 25 wt %; T_{250} <1426°C (2600°F)
Coal volatility	>30% dry basis	Less stringent	Less stringent

Source: [15]

Numerous other organizations also had significant initiatives in the CWS fuels industry. These included (not exhaustive) Babcock & Wilcox, Atlantic Richfield, U.S. Fluidcarbon (a subsidiary of Allis Chalmers, Ashland Oil Co., and AMAX. Of these, OXCE and Carbogel are used as examples of high density CWS fuels in subsequent discussions. Table 3.2 presents the fuel characterizations for OXCE and Carbogel high-density CWS fuels based upon eastern bituminous coals. High density CWS fuels have also been proposed as diesel fuel substitutes in Alaska, based upon the significant coal resource base in that locale. Table 3.3 presents characteristics of the coals proposed as the basis for a diesel fuel substitute. The CWS proposed for this location was based upon the subbituminous coals produced there. The CWS proposed

Table 3.2. Representative CWS Fuels Proposed for Oil Substitution

	Fuel			
CWS Producer	OXCE Fuel Co.		Carbogel	
Coal Supplier	Westmoreland Coal		Homer City	
Fuel Basis	As Rec'd	Dry	As Rec'd	Dry
Proximate Analysis (wt %)[2]				
Moisture	30.2	---	29.5	---
Volatiles	25.7	36.9	19.7	27.9
Fixed Carbon	39.8	57.0	46.0	65.2
Ash	4.3	6.1	4.9	6.9
Ultimate Analysis (wt %)[1]				
Carbon	55.5	79.5	56.8	80.6
Hydrogen	3.8	5.5	3.7	5.2
Oxygen	4.7	6.8	3.2	4.5
Nitrogen	1.0	1.4	1.1	1.6
Sulfur	0.6	0.8	0.8	1.2
Moisture	30.2	---	29.5	---
Ash	4.3	6.1	4.9	6.9
Higher Heating Value				
MJ/kg	23.12	33.09	21.66	32.50
Btu/lb	9,923	14,216	9,303	13,196
CWS Characteristics				
Solids Loading	69.8%		70.5%	
Viscosity, Pa·s (cp@100/s)	0.73 (725)		1.20 (1200)	

Notes: [1]Totals may not add to 100.0 due to rounding. Source: [17]

Table 3.3. Characteristics of Alaskan Coals Proposed for CWS Production

	Coals			
	Beluga Coal		Usibelli Coal	
	Eq. Moisture	Dry	Eq. Moisture	Dry
Proximate Analysis (wt %)				
Moisture	25.0	---	24.1	---
Volatiles	34.4	45.9	36.1	47.6
Fixed Carbon	35.6	47.4	32.2	42.4
Ash	5.0	6.7	7.6	10.0
Ultimate Analysis (wt %)				
Carbon	47.8	63.7	47.9	63.1
Hydrogen	6.4	4.8	6.4	4.9
Oxygen	40.0	23.7	37.3	20.9
Nitrogen	0.6	0.8	0.7	0.9
Sulfur	0.2	0.3	0.2	0.2
Moisture	25.0	---	24.1	---
Ash	5.0	6.7	7.6	10.0
Higher Heating Value				
MJ/kg	18.97	25.31	18.86	24.84
Btu/lb	8,150	10,870	8,100	10,670

Source: [12]

for Alaska was a low-sulfur fuel, resulting from the particular characteristics of Beluga and Nenana field coals.

Developing a high density CWS from Alaska subbituminous coals would have the consequence of producing a fuel with low energy density; however the sulfur and nitrogen contents of Alaskan coals is sufficiently attractive to overcome some of the problems associated with the low energy density. The project proposed for Alaska was to use waste coal fines from the Usibelli coal mine, prepare them with hydrothermal treatment to reduce the equilibrium moisture content by half and upgrade the heat content, and then produce the slurry. The CWS would be used initially as diesel fuel but ultimately would be shipped to Japan as a heavy fuel oil substitute [17].

Note that in all of these cases the energy density of the high-density CWS is no more than half of that typically associated with heavy fuel oils. The OXCE Fuel and Carbogel CWS products are typically 21.7 – 23.1 MJ/kg (9,100 – 9,300 Btu/lb). The CWS fuels reviewed by Babcock & Wilcox [12] have typical heat contents of 23.1 – 26.5 MJ/kg (9,900 – 11,400 Btu/lb). These compare to #6 heavy fuel oil at 40.6 – 44.2 MJ/kg (17,440 – 18,990 Btu/lb) [12]. The attractiveness of such high-density CWS products are the potential fuel cost, resource base, and energy security advantages for nations that are significant importers of petroleum.

While high-density CWS technology has not been deployed commercially in the USA, it has achieved commercial status as an approach to opportunity fuel production in other nations. A medium sized CWS project has been developed in Oristano Harbor, Sardinia, Italy [18]. This plant involves receiving coal, crushing it to <6 mm prior to storage in silos. Coal reclaimed from the silos is crushed to 0.5 mm for froth flotation, and then reground to <0.2 mm. About 33 percent of the coal is further ground to an ultra fine consistency. Following final grinding, a filter cake is produced. The filter cake is then converted into a slurry of ~70 weight percent solids with a viscosity of about 1 Pa·s for use in a local power plant, and for transport to other state-owned power plants in Italy [18].

Other nations have also performed considerable research on high-density CWS fuels including Japan, China, and Russia [17]. Trass [19], in a review of CWS technology, cites such developments as the Italian Porto Torres project with a capacity of 500,000 tonnes/yr, engineering and development of a 3 million tonne/yr project in Russia, including a 256 km CWS pipeline from the point of production to the point of use, and added research in Japan and China. A 500,000 tonne/yr project has been built and operated in Japan, and CWS has been exported from China to Japan [19 - 23].

China has initiated a significant program for the development of high-density CWS fuels as a diesel oil substitute [24-25]. In response to high oil costs in that developing economy, the World Laboratory in China is employing ultra-clean (low ash) micronized coal in a high-density coal-oil-water slurry to reduce fuel costs in agriculture and in rural areas. Millions of small diesel engines exist throughout China and are used in light tractors, harvesters, transportation systems, and water pumps supporting agriculture [24]. The high density CWOS produced experimentally was shown to be an adequate substitute for diesel oil, with only a slight degradation in engine performance [24].

China also houses one of the largest installations using high density CWS—the Maoming Thermal Power Plant [26]. Two 61.1 kg/s (485,000 lb/h) anthracite boilers were installed at this location in 1961, and these were retrofitted to fire oil in 1968. China, however, has few oil resources and significant coal resources; oil is expensive in that country. As a consequence, boiler #1 was retrofitted to fire high-density CWS in the year 2000, and boiler #2 was retrofitted to fire high-density CWS in 2001 [26]. Table 3.4 compares the properties of the CWS as-fired to the oil it replaced. Table 3.5 compares the performance of these boilers on high-density CWS and oil.

Table 3.4. Fuel Properties of Coal-Water Slurry and Oil at the Maoming Power Plant

Parameter	Coal-Water Slurry	Oil
Ultimate Analysis (wt %)		
Carbon	50.17	86.43
Hydrogen	3.22	10.59
Oxygen	5.67	0.75
Nitrogen	0.96	0.97
Sulfur	0.27	0.46
Moisture	34.29	0.80
Ash	5.42	0.00
Higher Heating Value (MJ/kg)	18.89	40.05
Higher Heating Value (Btu/lb)	8115	17.20
Viscosity (Pa·s)	1.18	---
Viscosity (centipoise)	1180	---

Source: [26]

Table 3.5. Critical Performance Parameters at the Maoming Power Plant When Firing CWS or Oil (values per boiler)

Parameter	Firing CWS	Firing Oil
Capacity (kg/s)	61.1	61.1
Capacity (lb/h)	485000	485000
Steam Pressure (MPa)	9.8	9.8
Steam Pressure (psia)	1420	1420
Steam Temperature (°C)	540	540
Steam Temperature (°F)	1005	1005
Furnace Exit Gas Temperature (°C)	953	1135
Furnace Exit Gas Temperature (°F)	1750	2075
Exhaust Gas Temperature (°C)	146	160
Exhaust Gas Temperature (°F)	295	320
Fuel Consumption (t/h)	15.1	31.0
Boiler Efficiency (%)	92.3	90.2

Source: [26]

Note that there is a modest reduction in the sulfur content of the fuel as fired when switching from oil to CWS. There is also a modest reduction in boiler efficiency caused by the higher moisture content of the fuel. The data in Table 3.5 clearly show that the Maoming boilers are quite capable of firing CWS without sacrificing boiler efficiency. Further, the boilers can operate at 40 – 100 percent of capacity with stable combustion when firing CWS at this location [26]. The conversion is quite successful because of the differences between the cost of oil and the cost of coal in China.

High density CWS programs developing alternatives to oil and oil-based fuels provide a significant opportunity fuel to achieve a measure of energy security in an uncertain world. They are particularly promising where coal and coal wastes are inexpensive, and where oil is an expensive commodity.

3.2.2. Characteristics of Low-Density Coal Water Slurry Fuels

While high-density CWS fuels, developed as a complete substitute for petroleum, have been subject to the whims of oil pricing [19], low-density (e.g., 45 – 55 weight percent solids) CWS fuels have been developed to address additional concerns: environmental and safety issues in coal mining and coal beneficiation, clean-up of waste coal ponds and impoundments where high concentrations of fine and ultra-fine coal exists, and control of airborne emissions—specifically NO_x. Unlike the high-density CWS fuels which are meant as complete substitutes for oil, the low-density CWS fuels are meant to be cofired in pulverized coal and cyclone boilers.

The technology for low-density CWS production largely involves either mining old coal ponds and impoundments, or obtaining fines from coal cleaning plants now in operation. Pond fines can be used directly, or can be further beneficiated to reduce the ash content. The consequent fuel can be significantly variable.

Low density CWS technology is particularly beneficial in handling fines and producing a useful fuel. Fine coal, without significant concentrations of lump coal, does not readily flow well in the initial material handling aspects of power plant systems. It is prone to bridging in hoppers and other storage devices. When cold and wet, such fines are particularly problematical for coal handlers. Bridging and plugging handling issues are readily addressed by slurrying the fines.

3.2.2.1. Basic Characteristics of CWS Fuels.

In the large study for the Electric Power Research Institute (EPRI), The Energy Institute of the Pennsylvania State University characterized fines supplied from a variety of sources. Table 3.6 presents characterization of as-received and cleaned coal fines from Pennsylvania bituminous coal, on a dry basis. Table 3.6 illustrates the wide variety in quality that can be obtained depending upon whether the fines removed from the ponds or impoundments are re-beneficiated or used in an as-produced condition. If economically feasible, re-beneficiation produces a higher quality fuel, without sacrificing the energy content of the deposit. The energy recovery is significant.

Some of the variability shown in Table 3.6 is based upon the age of the coal fines impoundment. Earlier impoundments, based upon older coal cleaning technology, may contain more fuel, and higher value fuel. More recent technology leaves less coal behind in the form of fines. Other factors influencing this variability include the Hardgrove Grindability Index of the coal being cleaned, and its inherent friability.

Table 3.6. Comparison of Proximate and Ultimate Analysis for As-Received and Cleaned Coal Fines in Pennsylvania

	Coal Fines	
	As-Mined	Cleaned Coal Fines
Proximate Analysis (wt %, dry basis)		
Volatile matter	22.1	29.8
Fixed carbon	36.6	61.5
Ash	41.3	8.8
Ultimate Analysis (wt %, dry basis)		
Carbon	48.7	78.6
Hydrogen	3.4	4.7
Oxygen	4.6	1.4
Nitrogen	0.9	1.2
Sulfur	1.1	5.3
Ash	41.3	8.8
Higher Heating Value		
MJ/kg	19.52	30.80
Btu/lb	8,384	13,229
Yield(wt basis)		56%
Energy recovery		88%

Source: [15].

Table 3.7 presents characteristics of coal pond fines supplied by TVA from eastern Alabama, pond fines supplied by Central Illinois Public Service from a local source, and fines from the coal cleaning plant at Rend Lake, IL. Table 3.7 illustrates the inherent variability depending upon location, coal cleaning approach taken originally, and related variables. Note that Table 3.7 also presents the particle size distribution of the fines produced, and the pH of the coal supply. The particle size distribution is particularly significant since it influences the viscosity of the resulting CWS fuel.

Apparent viscosity of three CWS fuels is presented in Figure 3.1. Note that the desired specification for low-density CWS is for an apparent viscosity of 0.2 Pa·s (200 cp @ 100/s), a condition which limits the solids loading in these low-density CWS fuels.

Table 3.7. Characteristics of Coal Fines for Possible CWS Production from 3 Sources (Analyses on dry wt basis)

	Source		
	Eastern Alabama	Central Illinois[1]	Rend Lake, IL
Proximate Analysis			
Volatile matter	20.7	24.8	35.5
Fixed carbon	60.4	44.6	58.5
Ash	18.9	30.5	6.0
Ultimate Analysis			
Carbon	71.6	55.7	76.2
Hydrogen	3.8	3.3	4.9
Oxygen	1.1	0.9	1.8
Nitrogen	0.6	2.0	1.0
Sulfur	4.0	7.6	10.1
Ash	18.9	30.5	6.0
Higher Heating Value			
MJ/kg	28.39	22.45	31.41
Btu/lb	12,195	9,642	13,493
PH	6.7	ND	4.5
Particle size distribution (μm)			
D (v, 0.9)	82	117	141
D (v, 0.5)	28	31	48
D (v, 0.1)	7	7	10

Note: [1]Total does not add to 100.0 due to rounding. Source [15].

The coal fines shown in Figure 3.1 are Monongahela fines from western Pennsylvania, fines from Central Illinois Power Service, and fines from TVA, obtained from eastern Alabama. The western PA and central IL fines exhibited essentially the same behavior. The fines from eastern AL produced a slurry with significantly higher viscosity for any given solids loading. The approximate equations describing the relationships between solids loading and viscosity are as follows [15]:

Monongahela fines:

$$V = 2x10^{-7} \text{ x } e^{0.2583S} \qquad [3\text{-}1]$$

Where V is viscosity (Pa·s) and S is solids loading (wt %).

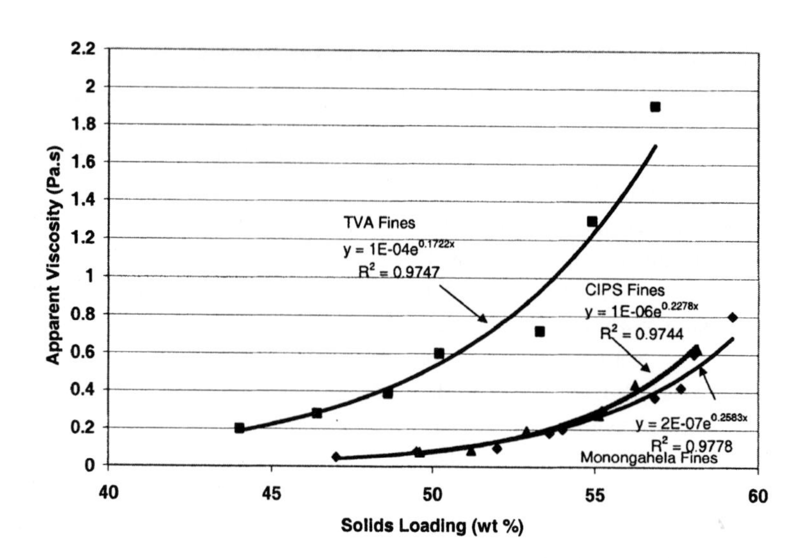

Figure 3.1. Apparent Viscosity of CWS Fuels from Selected Fines as a Function of Solids Loading Source: [adapted from 15]

Alternatively,

$$V_{cp} = 0.0073(S^2) - 0.7232(S) + 17.948 \qquad [3\text{-}2]$$

Where V_{cp} is viscosity in centipoise (100/s).

Central Illinois Public Service Fines:

$$V = 1x10^{-6} \text{ x } e^{0.2278S} \qquad [3\text{-}3]$$

Eastern Alabama Fines:

$$V = 1x10^{-4} \text{ x } e^{0.1722S} \qquad [3\text{-}4]$$

3.2.2.2. Reactivity and Environmental Characteristics of CWS Fuels.

The general position taken by researchers in the field is that the kinetics will mirror the parent coal—with some initial reaction suppression being caused by the water and ash content of the CWS [27]. At the same time reactivity measures can be calculated from the data presented previously in Tables 3.6 and 3.7. Such reactivity measures include volatile matter/fixed carbon (VM/FC) ratios from the proximate analysis, and atomic hydrogen/carbon (H/C) ratios and atomic oxygen/carbon (O/C) ratios from the ultimate analysis. Further, pollution measures can be calculated including kg S/GJ, kg SO_2/GJ, kg N/GJ, and kg ash/GJ.

It should be noted that certain reactions occur with bituminous coal-based CWS fuels. Coal particles in atomized droplets can agglomerate after the coal particles pass through their plastic stage, and the CWS fuels can produce char particles that are significantly larger than the starting coal particles. This increases the time required for char burnout, and has presented some problems for CWS combustion in boilers with tight furnace residence times and low furnace exit gas temperatures (FEGT) [27].

Reactivity measures are shown in Table 3.8 and pollution measures are shown in Table 3.9. It is important to note that Table 3.9 reflects the purpose of coal cleaning—removing ash and sulfur. The very high ash and sulfur concentrations particularly associated with the CIPS fines indicates the cleaning of high sulfur Interior Province coals.

Table 3.8. Reactivity of Several Coal Fines Used in CWS Preparations

	Source of Coal Fines				
Reactivity Measure	Western PA – as received	Western PA – cleaned	Eastern AL (TVA)	Central Illinois Pub Service (CIPS)	Rend Lake Cleaning Plant
VM/FC	0.604	0.485	0.343	0.556	0.607
FC/VM	1.656	2.604	2.918	1.798	1.648
H/C atomic	0.832	0.712	0.632	0.706	00.766
O/C atomic	0.071	0.013	0.012	0.012	0.018

Source: [15]

Table 3.9. Pollution Measures for Several Coal Fines Used in CWS Preparations

	Source of Coal Fines				
	Western PA – as received	Western PA – cleaned	Eastern AL (TVA)	CIPS	Rend Lake Cleaning Plant
kg S/GJ	0.564	1.723	1.410	3.389	3.219
lb S/10^6 Btu	1.312	4.006	3.280	7.882	7.485
kg SO_2/GJ	1.128	3.445	2.820	6.779	6.437
lb SO_2/10^6 Btu	2.624	8.013	6.560	15.764	14.971
kg N/GJ	0.462	0.390	0.212	0.892	0.319
lb N/10^6 Btu	1.073	0.907	0.492	2.074	0.741
kg ash/GJ	21.18	2.860	6.664	13.60	1.912
lb ash/10^6 Btu	49.26	6.652	15.498	31.632	4.447

Source: [15]

In addition to the pollution measures shown in Table 3.9, some attention must be given to trace metals in coal fines extracted from ponds and impoundments. Coal cleaning processes based upon specific gravity are designed to remove ash. There is a weak correlation between removing ash and removing trace minerals [28]. However Clarke and Sloss [28] show stronger relationships with specific metals and cleaning processes. Conventional coal cleaning removes 11 – 25 percent of the mercury and selenium in coal; 26 – 50 percent of the arsenic, cadmium, copper, nickel, and zinc in coal; and 51 – 70 percent of the lead in coal [28]. Consequently one could expect somewhat elevated concentrations of these metals in coal fines, relative to the parent coal, if the fines are not re-beneficiated. Advanced physical coal cleaning is shown to remove 11 – 25 percent of the copper in coal, 51-70 percent of the cadmium, nickel, and zinc in coal, and 71 – 90 percent of the arsenic and lead in coal [28]. Again concentrations of these metals may be elevated in coal fines extracted from ponds and impoundments.

Exact concentrations of trace metals in CWS fuels will vary dramatically depending upon the coal being cleaned, the cleaning process employed, and whether the fines are further cleaned before being used in a CWS formulation. Metals with high organic affinity or organic association such as beryllium, boron, titanium, and vanadium [29] may not report preferentially with the fines, and may not increase in concentration in the CWS fuel relative to the parent coal. Metal capture in the generation of coal fines creates a significant variable in the evaluation of sources of material for CWS fuels, due to the increasing attention and regulation of these pollutants [30].

3.2.2.3 Conclusions Regarding Low-Density CWS Fuels.

Low density CWS fuels, then, can provide a potentially low cost fuel for cofiring in utility boilers. The production of this fuel can be of environmental benefit to the coal preparation industry, reducing the accumulation of billions of tons of fines in ponds and impoundments. The fuels can have significant fuel content, and reactivity similar to the parent coals. Environmentally the use of these fuels can improve NO_x emissions, although they can increase SO_2 emissions depending upon the source of the fines. Firing high sulfur CWS product would require a flue gas desulfurization system or equivalent SO_2 capture system. Trace metals may increase in concentration depending upon the coal being processed, the coal cleaning technology, and the extent to which the fines are re-beneficiated prior to incorporation into a CWS fuel.

3.3. Case Studies in Cofiring Coal-Water Slurries

Two case studies firing CWS fuels have been extensively reported in the literature: cofiring at Seward Generating Station and cofiring at the Marion County Generating Station. These programs are summarized below.

3.3.1. Cofiring CWS at the Seward Generating Station

Seward Generating Station, under the ownership of GPU Genco, was the location of numerous opportunity fuel tests. It was the location of one of the most extensive development programs described in numerous papers and publications [13-14, 16, 31-34]. In the Seward CWS program, low-density slurries were prepared and fired in both Boiler #14 and Boiler #15. Boiler #14 was a 32 MW_e wall-fired Babcock& Wilcox boiler with

6 burners arranged in two rows of 3 burners on the front wall of the unit. It was installed in about 1950. Boiler #15, installed in 1957, was a 147 MWe tangentially-fired boiler equipped with four elevations of burners. Boiler #15 was equipped with a selective non-catalytic reduction (SNCR) system for NO_x reduction, based on urea injection. Of these cofiring test programs, the demonstration at Boiler #15 was of more interest and significance.

Working with The Energy Institute of The Pennsylvania State University and with EER Corporation, GPU Genco developed a CWS program. The specification for the CWS was a slurry containing about 50 percent solids (weight basis), and having an apparent viscosity of 0.2 Pa·s (200 centipoise @ 100/s). Cleaned fines were employed, with an ash content <10 percent, a sulfur content <1 percent, and a dry basis higher heating value of 32.6 MJ/kg (14,000 Btu/lb).

Fines were obtained from the Homer City preparation plant, from Washington Energy's processing plant near Pittsburgh, PA, and from the Russelton pond located in Allegheny County, PA. The Russelton site has over 5 million tons of fines from the mining of coal from the Upper Freeport seam. Table 3.10 summarizes the characteristics of these fines, and of the CWS fuel resulting from these fines. Note the comparison between the impounded fines and the CWS (dry basis) illustrating the benefit of additional fines cleaning.

The fines preparation system resulting in clean fines is shown in Figure 3.2. Once produced, the filtercake was transported to the Seward site and reformulated into the desired slurry defined in Table 3.10, and shown in Figure 3.3. The slurry was then fired in the boiler as is also shown in Figure 3.3. A moyno pump was used to transport the CWS.

Cofiring tests were conducted during the summer of 1997, with the plant operating at 130 – 136 MW_e. Cofiring percentages reached as high as 35 percent on a heat input basis. Operationally the unit experienced no particular difficulties. Attemperation sprays, burner tilts, and opacity did not change as a consequence of cofiring CWS in the boiler. Furnace exit gas temperatures decreased by about 11°C – 17°C (20°F – 30°F). There was a significant improvement in NO_x formation, however. This improvement was shown both as a function of reduced emissions and reduced urea consumption.

Table 3.10. Characteristics of Fuels Burned at Seward CWS Tests

	Fuel			
	Base coal	Coal-Water Slurry		
		Impounded coal fines	CWS (as received)	CWS (dry basis)
Proximate Analysis (Wt %)				
Moisture	1.6	N/A	51.2	---
Ash	14.2	41.3	4.8	9.9
Volatiles	22.6	22.1	14.3	29.3
Fixed Carbon	63.2	36.6	29.7	60.8
Ultimate Analysis (Wt %)				
Carbon	74.6	48.7	37.9	77.6
Hydrogen	4.3	3.4	2.3	4.7
Oxygen	3.9	4.6	2.5	5.2
Nitrogen	1.3	0.9	0.6	1.3
Sulfur	1.7	1.1	0.7	1.4
Moisture	1.6		51.2	---
Ash	14.2	41.3	4.8	9.9
Higher Heating Value				
MJ/kg	29.65	17.59	15.32	31.40
Btu/lb	12737	7554	6582	13487
CWS Characteristics				
Solids loading	N/A	N/A	48.8	48.8
Viscosity (Pa·s)			0.1	0.1
Viscosity (cp @ 100/s)	N/A	N/A	99	99
PH	N/A	N/A	6.5	6.5
Particle size distribution (μm)				
99.8% passing	N/A	N/A	N/A	293
D(v, 0.9)	N/A	N/A	N/A	142
D(v, 0.5)	N/A	N/A	N/A	30
D(v, 0.1)	N/A	N/A	N/A	6

Source: [33]

The NO_x benefit can readily be seen in Figures 3.4 and 3.5. Initial testing consisted of setting the NO_x emissions at 0.22 kg/GJ (0.51 lb/10^6 Btu) and tracking the consumption of urea to achieve that emission level. Urea consumption declined from 5.7 – 6.9 l/s (90 – 110 gal/hr) to

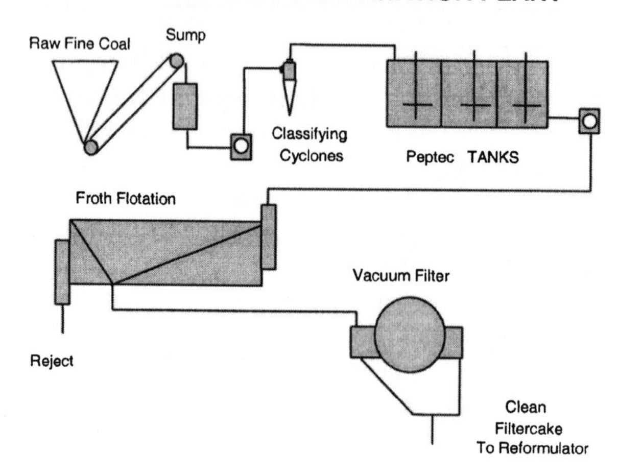

Figure 3.2. The Fines Preparation System at Washington Energy Processing
Source: [14]. (Used with permission)

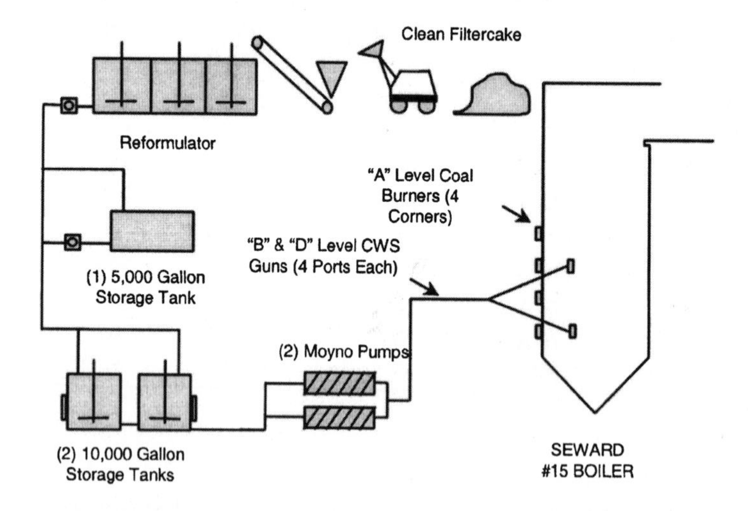

Figure 3.3. The CWS Reformulation and Firing System at Seward Generating Station.
Source: [14]. (Used with Permission)

about 3.2 l/s (about 50 gal/h). This was achieved while firing 4.1 l/s (65 gpm) of CWS—far below the capacity of the system. Figure 3.4 shows that, by firing 6.6 l/s (104 gpm) of CWS, the SNCR system could be turned off—no urea was required—and the NO_x emissions were maintained at about 0.22 kg/GJ (0.5 lb/10^6 Btu).

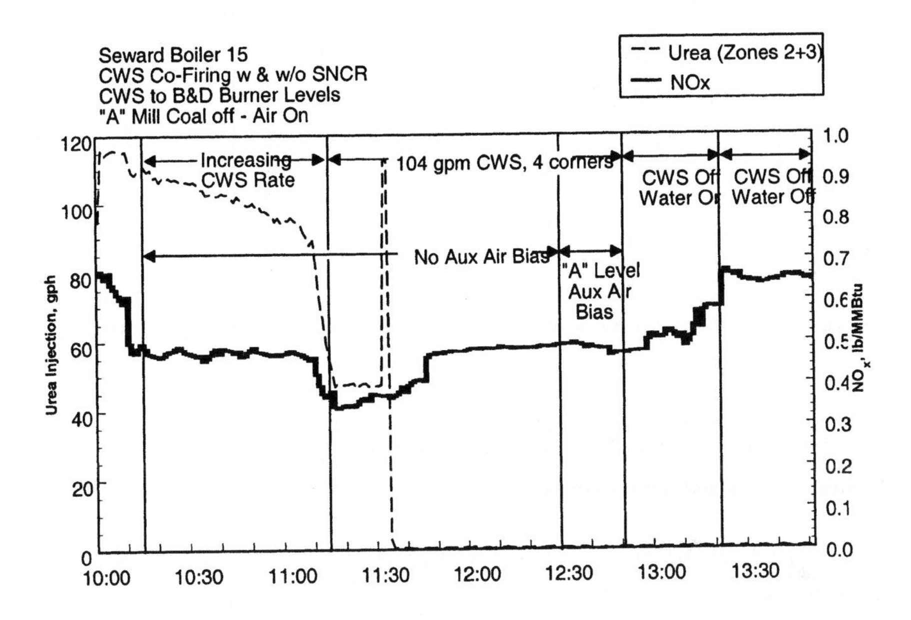

Figure 3.4. NOx Emissions at Seward Generating Station Cofiring 393 l/m CWS and with the SNCR Shut Off

Source: [14]. (Used with permission)

The advantage of CWS can be taken either in the form of reduced NO_x emissions—if such emissions are close enough to regulatory requirements to avoid investment in a post-combustion device. Alternatively the benefits can be taken in reduced consumption of urea or ammonia.

It is significant that the cofiring of CWS to achieve NO_x emissions reductions goes beyond the use of this as a water injection technique. Figure 3.5 shows the reduction in urea consumption when cofiring 7.6 l/s (121 gpm) of CWS with NO_x reduced to 0.155 kg/GJ

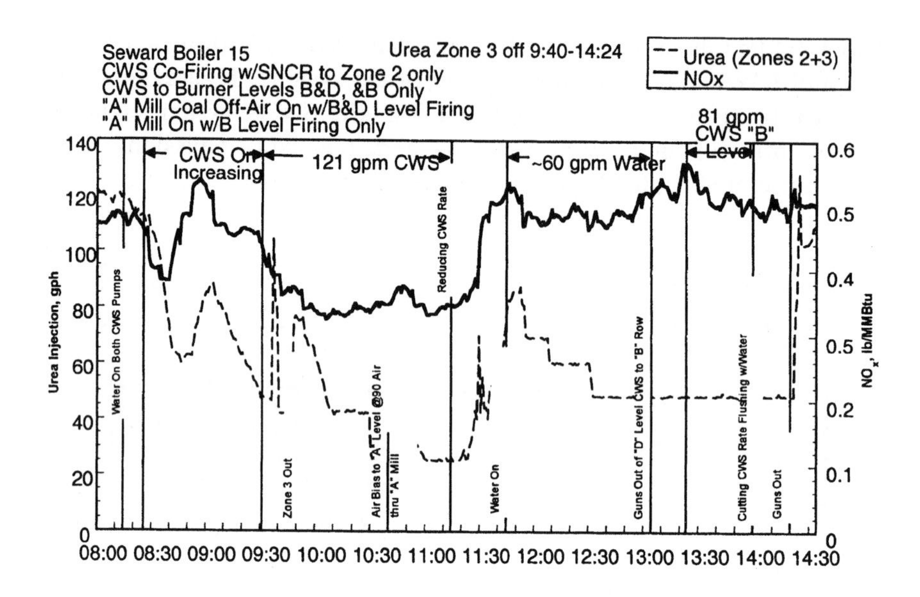

Figure 3.5. The Influence of CWS Cofiring, as Opposed to Equivalent Water Injection, on NOx Emissions at Seward Generating Station
Source: [14] (Used with permission)

(0.36 lb/10^6 Btu). When 3.8 l/s (60 gpm) of water was injected—an amount equivalent to the water in the CWS—and the urea was held reasonably constant, NO_x emissions again increased.

The extensive testing conducted at Seward Generating Station was directed towards the NO_x benefit, recognizing that CWS would cause a small efficiency penalty but no operating difficulties. This testing demonstrated the significant advantages of the CWS fuel and, through specification, managed the issues of SO_2 control. These tests proved the concept of CWS cofiring. However the station was sold from GPU to Sithe Energies, and then to Reliant Energy. It is now being repowered.

3.3.2. Cofiring CWS at Marion, Illinois

Cofiring CWS at Boiler #3 of Southern Illinois Power Cooperative (SIPC) Marion, IL documented the potential for using CWS fuels in cyclone boilers. This program followed a short demonstration of

CWS cofiring in a single cyclone barrel at the Paradise Fossil Plant of TVA. The testing at Paradise Fossil Plant occurred in a single barrel on Unit #1, a 700 MW_e unit with 14 cyclone barrels arranged in an opposed firing configuration. Fines used in the formulation of CWS came from the coal preparation plant on the Paradise site. The TVA test demonstrated the technical feasibility of cofiring CWS in cyclone barrels. It showed that cofiring in the range of 10 – 30 percent on a heat input basis caused no operational difficulties [35].

Testing of cofiring impacts on a cyclone boiler then proceeded at the Marion, IL location. Boiler #3 at the Marion Power Plant is a 33 MW_e unit fired with two 2.15 m (7 ft) diameter cyclone barrels. Coal pond fines were windrow dried and then fed to a high-energy mixer. The product was then stored in a storage tank prior to firing. When recovered from the tank, the slurry was pumped at rates up to 2.8 l/s (45 gpm) into the two cyclones. This achieved a cofiring rate of 44 percent on a heat input basis. The base fuel during these tests had a heating value of 30.67 MJ/kg (12,885 Btu/lb), a sulfur content of 3.32 percent, and an ash content of 9.96 percent [35]. Table 3.11 shows the ultimate analysis of the CWS fired during these tests, conducted Sep 30 – Oct 4, 1996. Note from the moisture contents that solids loadings ranged from 43.4 percent to 56.8 percent.

Table 3.11. Ultimate Analyses of Coal Water Slurries Burned at Southern Illinois Power Cooperative Marion Station Unit #3

Sample Date	9/30/96	10/1/96	10/2/96	10/3/96	10/4/96
Ultimate Analysis (wt %, as-received)					
Carbon	25.87	28.39	29.45	29.96	30.98
Hydrogen	1.51	1.56	1.62	1.75	1.73
Oxygen	3.62	5.17	4.55	4.51	4.39
Nitrogen	0.54	0.65	0.63	0.66	0.67
Sulfur	1.03	0.71	0.80	1.00	1.02
Moisture	56.62	52.13	50.01	47.41	43.25
Ash	10.81	11.38	12.95	14.71	17.9
Higher Heating Value					
MJ/kg	10.71	11.52	11.68	12.08	12.25
Btu/lb	4,601	4,949	5,016	5,188	5,264

Source: [35]

The testing at Marion, IL, was conducted by EER Corporation (now GE-EER), using its own burner/injector design. Operationally the tests at Marion, IL demonstrated a practical cofiring limit of 20 percent on a heat input basis. Above that level both opacity and carbon conversion efficiency performance experienced degradation. Flyash carryover increased markedly. It should be noted, however, that the fines used in the CWS product at Marion, IL were coarser than those used at Seward Generating Station. Cofiring CWS at the SIPC Boiler #3 decreased cyclone barrel temperatures by 50°C – 90°C (90°F – 160°F), however there were no reports of slag freezing in the cyclone barrels. That was consistent with the operating experience at Paradise Fossil Plant. From a boiler capacity and operability perspective, then, cofiring was not of consequence at this installation.

Cofiring percentages at SIPC Boiler #3 reached 44 percent on a heat input basis. The best NO_x reduction occurred at 34 percent cofiring (heat input basis), when NO_x levels declined by 11 percent. Again the NO_x reductions achieved exceeded those associated with water injection alone [35]. The lack of significant NOx benefit was attributed to the low excess O_2 used in firing the system, and the consequent low NO_x emissions for cyclone boiler firing—typically 0.41 kg/GJ (0.96 lb/10^6 Btu) at this location when firing at full load.

3.4. Opportunity Fuels Related to CWS Technology

In addition to the CWS fuels described above, numerous organizations have developed variations on the CWS technology to utilize waste coal fines. The National Energy Technology Laboratory of USDOE developed and patented a technology called GranuFlow™ to increase utilization of waste coal fines. In this technology, fines are agglomerated with a small amount of bitumen or heavy oil emulsion. The petroleum product is added to fine coal before it is dewatered to produce a product that is more readily handled [10, 36]. This technology has been licensed to CQ Inc. of Homer City, PA for commercialization. CQ Inc has demonstrated that the fines can be more completely dewatered by 2 – 6 percent, and that the product is more readily handled. Orimulsion™ has been used successfully as the binder source for producing GranuFlow™. Bitumen from oil refining also has been used.

CQ Inc. has a tradition of developing waste-based fuels, including E-fuel™, a pelletized product comprised of coal fines and pulp mill sludge. E-fuel™ pellets are used as stoker fuel where the combination of coal fines and waste binder is economically available [37]. This firm has extended the GranuFlow™ technology by blending pulp mill sludge with coal fines and a bitumen binder [37-38]. This product, which has potential if blended with coal, recently was tested at the R. Paul Smith Generating Station of Allegheny Energy Supply Co., LLC. in a blend with low sulfur coal.

The Center for Applied Energy Research of the University of Kentucky has performed additional research on producing usable fuel from coal fines, dewatering them in a substrate of plastic, newsprint, carpet wastes, mixed office wastes, raw paper, and wood fibers [39]. This research holds promise for reducing the moisture content of waste coal fines and producing a useful fuel while holding the costs of dewatering down through the use of other waste products.

Dooher and Lebowitz [40] of Adelphi University have investigated making CWS products using non-dewatered sewage sludge as the source of liquid. Florida Power Corporation commissioned Adelphi to examine the feasibility of this approach to CWS production. The process envisioned includes hydrothermal treatment of the sewage sludge to improve the properties of the CWS. While this technology has not progressed to demonstration or commercialization stages, it exemplifies the creativity brought to bear on the use of waste coal fines. Antaeus Energy is investigating a means for recovering coal pond fines in WV, and producing a coke product for the steel industry [41]. This represents yet another innovative approach to recovering coal pond fines as an opportunity fuel.

3.5. Conclusions Regarding Coal-Water Slurries

The 2 billion tons of waste coal fines existing in impoundments and ponds is an environmental and safety problem for the coal mining and coal preparation community. Coal fines do not readily dewater, and will remain in suspension in water—the primary transport medium in the coal cleaning processes—for long times. If stocked out and dried, these fines become a fugitive dust emissions source. More likely, however, they will be handled by being placed behind the many dams in impoundments developed for such materials. In 1972, the Buffalo Creek, WV impoundment failed killing 125 people, injuring over 1,000 additional persons, and leaving more than 4,000 individuals homeless [42-44]. This disaster destroyed 502 homes and 44 mobile homes, and did damage to an additional 270 homes along Buffalo Creek. It also destroyed about 1,000 automobiles and trucks, some highway and railroad bridges, sections of highway, electricity transmission and distribution lines, telephone lines, and other infrastructural items [42].

More recently, in October 2000, the Martin County Coal Corporation's Big Branch Refuse Impoundment near Inez, KY failed, releasing 1.16×10^6 m^3 (306×10^6 gal) coal fines through the active mine works of the 1-C Mine. The Inez, KY impoundment failure did not cause any deaths, but it did contaminate the Wolf Creek and Coldwater Fork, the Big Sandy River, and the Ohio River with coal fines, closing several water treatment facilities. The contamination extended over 100 miles downstream from the mine and the Impoundment [45]. With over 2300 coal fines impoundments, of which some 1500 are abandoned, more impoundment failures can be expected in the future. Since 1972 there has been at least one failure or partial failure annually [44].

The production of CWS fuels from waste coal fines provides a way to address these environmental and safety problems. CWS fuels can be developed at high densities, as complete substitutes for heavy oil. Alternatively they can be developed at lower densities as an opportunity fuel to be cofired in large utility coal-fired boilers. The fuels can also be configured for cement kilns and, if pelletized, for stokers. CWS technology provides a mechanism for producing low cost fuel supplies. For many, it provides a means to increase energy security. That may explain the international appeal for developing CWS technology.

While CWS fuels address one of the mine safety and environmental issues, they also can be used to address airborne emissions from coal-fired boilers. Specifically cofiring CWS in coal-fired boilers can be used as a significant method for trimming NO_x emissions and reducing the costs associated with NO_x control. The extent to which CWS improves NO_x emissions can be a function of the base fuel and base combustion conditions, the slurry formulation, and the approach to slurry injection. The experimentation at Seward Generating Station Boiler #15, in particular, demonstrates the environmental potential of this family of opportunity fuels.

3.6. Utilization of Waste Coal as Opportunity Fuel

Anthracite culm and bituminous gob, by themselves, are also used as opportunity fuels. With the commercialization of circulating fluidized bed (CFB) technology, these waste coals became a low cost source of energy in traditional mining areas such as the anthracite fields of eastern PA, and the bituminous coal regions of western PA and WV. Other projects have been built in such locations as Hunosa, Austurias, Spain [46].

With the advent of CFB technology and its successful application to this fuel, there has been a significant growth in the use of anthracite culm and bituminous gob within the USA, as is shown in Table 3.12. This growth is built upon several factors that have coalesced: 1) the availability of hundreds of millions if not billions of tons of culm and gob, 2) laws and regulations that facilitate—perhaps promote—the development of independent power plants selling electricity to the grid, and 3) the ability of utilities to invest outside their traditional territories through non-regulated subsidiaries [47].

Table 3.12. Waste Coal Consumption in the USA by Year

Year	Quantity		Energy Content[1]	
	10^6 tonnes	10^6 tons	PJ	10^{12} Btu
1989	1.3	1.4	17.7	16.8
1990	1.9	2.1	26.6	25.2
1991	3.6	4.0	50.6	48.0
1992	5.7	6.3	79.7	75.6
1993	7.4	8.1	107.6	97.2
1994	7.5	8.2	102.3	97.1
1995	7.8	8.6	108.8	103.2
1996	8.0	8.8	111.3	105.6
1997	7.4	8.1	107.6	97.2
1998	7.9	8.7	110.1	104.4
1999	7.9	8.7	110.1	104.4
2000	8.3	9.1	115.1	109.2
2001	8.5	9.3	117.6	111.6

Note: [1]Assumes 6,000 Btu/lb on average for waste coal

Source: [48]

The growth of use in waste coal has occurred most significantly in the state of PA, where sufficient numbers of projects exist to support an association: The Anthracite Region Independent Power Producers Association (ARIPPA). This association includes both anthracite culm and bituminous gob projects in the state. Its member projects are shown in Table 3.13. In addition to the ARIPPA membership, which represents 1,343 MW_e of power plus considerable energy supplied in cogeneration or combined heat and power (CHP) applications, bituminous gob is used at the North Branch Power Partners project in Grant County, WV and the American Bituminous Power Partners project in Grant Town, WV. With the exception of the Seward Generation Station repowering project shown in Table 3.13, these projects are typically small, with typical capacities being <100 MW_e. Typical steam flows are in the range of 45 – 100 kg/s (355 – 795 x 10^3 lb/h) and typical steam conditions are on the order of 105 bar/513°C for non-reheat units, and 174.4 bar/540°C/540°C (2530 psig/1005°F/1005°F) for the reheat units [46].

Table 3.13. Anthracite Region Independent Power Producers Association Plants

Plant	Fuel	Capacity
Cambria Cogen Company	Bituminous coal waste (gob)	85 MW_e
Ebensburg Power Company	Bituminous coal waste (gob)	47 MW_e
Gilberton Power Company	Anthracite refuse (culm)	80 MW_e
Inter-Power/AhlCon Partners LP	Bituminous coal waste (gob)	102 MW_e
Mt. Carmel Cogeneration, Inc.	Anthracite refuse (culm)	40 MW_e
Northeastern Power Company	Anthracite refuse (culm)	50 MW_e
Panther Creek Partners	Anthracite refuse (culm)	80 MW_e
PG& E Northampton Project	Anthracite refuse (culm)	98 MW_e
PG& E Scrubgrass Project	Bituminous coal waste (gob)	87 MW_e
Piney Creek LP	Bituminous coal waste (gob)	32 MW_e
Reliant Energy–Seward Station	Bituminous coal waste (gob)	520 MW_e[1]
Schuylkill Energy Resources	Anthracite refuse (culm)	80 MW_e
Wheelabrator Frackville Energy Company	Anthracite refuse (culm)	42 MW_e

Note: [1] Projected for completion in May 2004

Source: Anthracite Region Independent Power Producers.

Waste coal utilization is not restricted to the northeastern US; it is occurring throughout the world. Further, it is being proposed and used in unconventional ways. The Healy, AK Clean Coal Slagging Combustor project is a 50 MWe generating station based upon the TRW slagging combustion technology. It is fueled with a blend of 50 percent run-of-mine Usibelli coal and 50 percent waste coal [49]. Waste Management and Processors, Inc. (WMPI) will expand the use of anthracite culm at the Gilberton site, constructing an indirect liquefaction facility there to produce diesel fuel. The culm will be gasified to produce a synthesis gas; the synthesis gas will then be converted to diesel fuel.

3.6.1. Characteristics of Waste Coals

Waste coals vary dramatically as a function of the coal seam, the effectiveness of the breaker boys in removing slate, and the extent to which the original culm or gob pile was remined. Table 3.14 illustrates the variability in anthracite culm, based upon two PA projects.

Table 3.14. Characteristics of Anthracite Culm Burned at two Projects

Parameter	Mt. Carmel – culm	Northampton - culm
Proximate Analysis (wt %)		
Moisture	10.0	8.15
Ash	63.63	38.67
Volatile Matter	---	6.87
Fixed Carbon	---	46.31
Ultimate Analysis (wt %)		
Carbon	20.62	45.92
Hydrogen	0.84	1.52
Oxygen	4.13	4.83
Nitrogen	0.38	0.34
Sulfur	0.40	0.56
Moisture	10.0	8.15
Ash	63.63	38.67
Higher Heating Value		
MJ/kg	7.57	16.86
Btu/lb	3,250	7,249

Source: [46].

Note that this is a fuel with very low reactivity, even by anthracite standards. Further, both fuels are very high in ash content. Some of the projects employ coal cleaning to improve the quality of the anthracite culm, with significant results. Table 3.15 demonstrates the utility of this practice at the Gilberton site.

Table 3.15. Ultimate Analysis of Unprocessed and Processed Anthracite Culm Burned at the Gilberton Power Company

Ultimate Analysis	Unprocessed	Processed
Carbon	26.82	48.25
Hydrogen	1.05	1.59
Oxygen	3.02	2.92
Nitrogen	0.50	0.88
Sulfur	0.25	0.76
Moisture	8.98	15.05
Ash	59.38	30.55
Higher Heating Value		
MJ/kg	10.1	18.1
Btu/lb	4,338	7,750

Source: [46].

Reprocessing the culm product, while increasing the moisture content, reduces the ash content by half. This significantly increases the heating value of the fuel. As noted previously, these are quite unreactive fuels. However, with the exception of ash content, they do not exhibit severe environmental problems. Table 3.16 presents reactivity and pollution measures for the anthracite culm fuels shown above.

Bituminous gob is also quite variable. A representative characterization of this material is shown in Table 3.17, based upon the gob burned at the Colver, PA plant. Note that gob, like its parent coal, is inherently considerably more reactive than anthracite culm as shown by its VM/FC ratio and its H/C and O/C atomic ratios. At the same time the ash content is comparable to that of the anthracite culm piles.

Table 3.16. Reactivity and Pollution Measures for some Anthracite Culm

	Fuel Source			
	Mt. Carmel	Northampton	Gilberton - unprocessed	Gilberton - processed
Reactivity				
H/C Atomic	0.485	0.394	0.466	0.393
O/C Atomic	0.150	0.079	0.084	0.045
Pollution				
Sulfur kg/GJ	0.529	0.332	0.248	0.422
Sulfur lb/10^6 Btu	1.231	0.773	0.576	0.981
SO_2 kg/GJ	1.059	0.665	0.495	0.844
SO_2 lb/10^6 Btu	2.462	1.545	1.153	1.961
Nitrogen kg/GJ	0.503	0.202	0.495	0.581
Nitrogen lb/10^6 Btu	1.169	0.469	1.153	1.135

Source: [46]

Table 3.17. Characteristics of Bituminous Gob Burned at the Colver Plant

Parameter	Value
Proximate Analysis (wt %)	
Moisture	5.04
Ash	38.52
Volatile Matter	17.47
Fixed Carbon	38.97
Total	100.0
Ultimate Analysis (wt %)	
Carbon	46.49
Hydrogen	2.53
Oxygen	4.27
Nitrogen	0.85
Sulfur	2.21
Moisture	5.04
Ash	38.97
Total	100.0
Higher Heating Value	
MJ/kg	19.17
Btu/lb	8240

Source: [46].

3.6.2 A Case Study in the Combustion of Waste Coal

The Northampton anthracite culm-burning plant shown in Figure 3.6 has been chosen for a case study. Numerous case studies could be used to represent the combustion of waste coals, including the Panther Creek project described by Svendsen and LeClerc [50] or the award winning Ebensburg Power Company plant [51]. The Northampton, PA project, however, represents the coalescing of all elements in waste coal combustion and is used here to represent the type of project typically developed to use this opportunity fuel.

The Northampton plant was built as a "Brown field" redevelopment project on the site of the former Universal Atlas Cement Company. The project is owned by Northampton Generating Company, LP, an affiliate of US Generating Company. The plant generates 100 kg/s of 174 bar/540°C/540°C (795 x 10^3 lb/h of 2530 psig/1005°F/1005°F) steam for the generation of 110 MW_e of electricity. The plant also supplies steam to a paper company.

Figure 3.6. The Northampton Anthracite Culm Burning Circulating Fluidized Bed Boiler

The plant, supplied by Foster Wheeler, burns nearly 500,000 tonnes (550,000 tons) of culm annually, and will consume over 14 million tonnes of this coal waste during its life. It is obtaining culm from 18 different locations, permitting redevelopment of that land as well as providing for the "brown field" redevelopment at the power plant site. Testing at the site has demonstrated that the Northampton facility, operating with a stoichiometric ratio of 1.25 and with an air heater exit temperature of 149°C (300°F) has a boiler efficiency of 87.9 percent [46] – which is comparable to the large utility steam generators fired with coal and natural gas.

Airborne emissions from the Northampton plant are shown in Table 3.18. These emissions are compared to the Colver plant, burning bituminous gob. Note that the plant meets very stringent environmental requirements. In achieving these levels the Northampton plant consumes 1.03 kg/s (4.14 ton/hr) of limestone. Because of the success of the plant, it received the Pennsylvania Governor's Award for Environmental Excellence in 1996. It also received the Job Creator Award from the Pennsylvania Employer Advisory Council [52].

Table 3.18. Airborne Emissions from the Northampton Plant, Compared to the Colver Bituminous Gob Plant

	Independent Power Plant	
	Northampton Plant	Colver Plant
NO_x, mg/MJ	41.18	58.5
SO_2, mg/MJ	50.98	132.75
CO, mg/MJ	24.44	35.55
HC, mg/MJ	0.18	0.45

Source: [46].

The ash in waste coal, as experienced in all of the plants identified above, is of critical importance. High ash contents can increase maintenance costs, particularly in crushers and coal handling systems. Further, the split between bottom ash and flyash must be carefully predicted such that solids removal does not become inordinately problematical.

3.7. Conclusions

Coal-water slurries and waste coals such as anthracite culm and bituminous gob provide significant opportunities to utilities, industries, and independent power producers. These opportunity fuels result from both past and present mining practices. They are typically high ash fuels; they may be high moisture fuels as well.

Coal-water slurries represent an untapped potential in the USA, and a developing industry in other parts of the industrial world. They can address significant environmental problems of the mining industry while reducing NO_x emissions from power plants. Waste coals represent a potential now being tapped wherever practical and economically attractive. For both types of material, proximity of the opportunity fuel to the point of use is essential. These fuels cannot be transported economically over long distances. For both types of material, creativity in the form and manner of use is also significant.

The development of these opportunity fuels is progressing based upon creativity in research on fuel formulation and fuel utilization. Of particular significance is the continual improvement in combustion technology—the development of cofiring technologies for CWS firing and the development of CFB technology for waste coal utilization.

3.8. References

1. Leonard, J.W. (ed). 1979. Coal Preparation. 4th Ed. American Institute of Mining, Metallurgical, and Petroleum Engineers. New York.
2. Anon. 2000. Coal Preparation. U.S. Department of Labor Mine Safety and Health Administration. Instruction Guide Series IG 54. Washington, D.C.
3. Speight, J.G. 1989. *The Chemistry and Technology of Coal.* Marcel Decker. New York.
4. Anon. 1994. State Coal Profile: Pennsylvania. Energy Information Administration, USDOE. Washington, DC.
5. Anon. History of Anthracite Coal Mining. Mine Safety & Health Administration, U.S. Department of Labor. Washington, DC.

6. Rhone, R.D. 1902. Anthracite coal mines and mining. American Monthly Review of Reviews. November.
7. Falcone Miller, S., J.L. Morrison, and A.W. Scaroni. 1995. The Utilization of Coal Pond Fines as Feedstock for Coal-Water Slurry Fuels. Proc. 20th International Technical Conference on Coal Utilization & Fuel Systems. Coal & Slurry Technology Association. Clearwater, FL. March 7-10.
8. Ashworth, R.A., T.A. Melick, D.K. Morrison, and J.J. Battista. 1998. Electric Utility CWS Firing Options to Reduce NO_x Emissions. Proc. 23rd International Technical Conference on Coal Utilization and Fuel Systems. Coal Technology Association. Clearwater, FL. March 9-13. pp. 719 – 730.
9. Mohanty, M.K, B.C. Paul, and R. Geilhuasen. 2003. Coal Preparation Plant Fine Waste: a Fuel Feedstock for Mine-Mouth Power Plants. Proc. 28th International Technical Conference on Coal Utilization & Fuel Systems. Coal Technology Association. Clearwater, FL. March 9-13.
10. Dilo Paul, A. and A. Deurbrouck. 2003. Drivers for Fine Coal Utilization. Proc. 28th International Technical Conference on Coal Utilization & Fuel Systems. Coal Technology Association. Clearwater, FL. March 9-13.
11. Trass, O. and E. Gandolfi. 1999. Coal-Slurry Fuels for Environmental Benefit. Proc. 24th International Technical Conference on Coal Utilization and Fuel Systems. Coal Technology Association. Clearwater, FL. March 8 - 11. pp. 823 - 832.
12. Stultz, S.C. and J.B. Kitto (eds). 1992. *Steam: Its Generation and Use.* 40th Ed. Babcock & Wilcox. Barberton, OH.
13. Morrison, J.L., B.G. Miller, and A.W. Scaroni. 1997. Coal-Water Slurry Fuel Production: Its Evolution and Current Status in the United States. Proc. 14th International Pittsburgh Coal Conference. Taiyuan, Shanxi Province, Peoples Republic of China. Sep 23-27.
14. Battista, J.J. 2002. Low Solids Coal Water Slurry Cofiring for NO_x Trim. Proc. 27th International Technical Conference on Coal Utilization and Fuel Systems. Coal Technology Association. Clearwater, FL. March 4-7. pp. 453 – 464.
15. Morrison, J.L., B.G. Miller, and A.W. Scaroni. 1998. Determining Coal Slurryability: A UCIG/Penn State Initiative. Electric Power Research Institute, Palo Alto, CA.

16. Morrison, J.L., B.G. Miller, A.W. Scaroni, and J.J. Battista. 1997. Coal Fines: A Discussion of their Utilization to Produce a Low Solids Coal-Water Slurry Fuel for Utility Cofiring Applications. Proc. Effects of Coal Quality on Power Plants. Electric Power Research Institute. Kansas May 20-22.
17. Wilson, W., C. Benson, R. Wilson, G. Krier, M. Ruckhaus, D. Walsh, and C. Ward. 1998. Alaska Low-Rank Coal-Water Fuel – Diesel Demonstration. Proc. 23rd International Technical Conference on Coal Utilization and Fuel Systems. Coal Technology Association. Clearwater, FL. March 9 - 13. pp. 1087 – 1097.
18. Bozano, L., S. Bozano, K. Holberg, P. Bozano, and G. Ferrara. 1999. Production of C.W.F. with Low Ash and High Percentage Solids. Proc. 24th International Technical Conference on Coal Utilization and Fuel Systems. Coal Technology Association. Clearwater, FL. March 8 - 11. pp.835 - 840.
19. Trass, O. 2001. Coal-Slurry Fuels: Where are We and Where Should We Go? Proc. 26th International Technical Conference on Coal Utilization and Fuel Systems. Coal Technology Association. Clearwater, FL. March 5 - 8. pp. 297 - 307.
20. Ercolani, D. and U. Tiberio. 1994. Start-up and Initial Operating Experience of Porto Torres Integrated Plant for Production and Utilization of Beneficiated Coal-Water Fuels. Proc. 19th International Technical Conference on Coal Utilization and Fuel Systems. Coal Slurry Technology Association. Clearwater, FL. pp. 9 – 20.
21. Ishibashi, Y., N. Abe, T.Kondo, and M. Gohta. 1994. Operating Experience of Large Scale CWM Production and Transportation. . Proc. 19th International Technical Conference on Coal Utilization and Fuel Systems. Coal Slurry Technology Association. Clearwater, FL. pp. 21 - 32.
22. Khodakov, C.S. The Statistical Analysis of the Pilot-Commercial Operational Results of the Power Generation and Delivery Complex Belovo-Novosibirsk. . Proc. 19th International Technical Conference on Coal Utilization and Fuel Systems. Coal Slurry Technology Association. Clearwater, FL. pp. 863 – 873.
23. Tanaka, M. 1996. The Operation of a CWM Relay Station in the CWM Chain from China to Japanese Users. . Proc. 21st International Technical Conference on Coal Utilization and Fuel Systems. Coal Slurry Technology Association. Clearwater, FL. pp. 691 - 698.

24. Zune, W., C. Baoming, and W. Xinwen. 2003. Rig Tests of a High-Speed Diesel Engine Fueled by Ultra-Clean Micronized Coal Slurry. Proc. 28th International Technical Conference on Coal Utilization & Fuel Systems. Coal Technology Association. Clearwater, FL. March 9-13.
25. Zune, W., F. Xiaoheng, and W. Xinguo. 2003. Preparation of Ultra-Clean Micronized Coal Slurryfuel as an Alternative to Diesel Oil. Proc. 28th International Technical Conference on Coal Utilization & Fuel Systems. Coal Technology Association. Clearwater, FL. March 9-13.
26. Xiang, Z., C. YuHaimiao, C. Xinyu, H. Zhenyu, L. Jianzhong, Z. Zhijun, Z. Junhu, and C. Kefa. 2003. The Application of Coal Water Slurry on the 220 t/h Utility Oil-Fired Boiler in Maoming Thermal Power Plant. Proc. 28th International Technical Conference on Coal Utilization & Fuel Systems. Coal Technology Association. Clearwater, FL. March 9-13.
27. Miller, B. 1993. Personal communication. July 15.
28. Clarke, L.B. and L.L. Sloss. 1992. Trace Element Emissions from Coal Combustion and Gasification. IEA Coal Research. London, England.
29. Tillman, D.A. 1994. *Trace Metals in Combustion Systems.* Academic Press. San Diego, CA.
30. Tillman, David A. 1998. *Trace Metal Emissions from Coal Combustion and Gasification*. Encyclopedia of Environmental Analysis and Remediation. Wiley Interscience. New York. pp. 4837-4847.
31. Battista, J.J., and R.A. Ashworth. 1998. Co-firing Coal Water slurry Fuel on a Tangentially-Fired Boiler. Proc. 23rd International Technical Conference on Coal Utilization and Fuel Systems. Coal Technology Association. Clearwater, FL. March 9 - 13. pp. 361 - 370.
32. Morrison, J.J., B.G. Miller, and J.J. Battista. 1998. Recovery of Ultra Fine Bituminous Coal from Screen-Bowl Centrifuge Effluent: A Possible Feedstock for Coal-Water Slurry Fuels? Proc. 23rd International Technical Conference on Coal Utilization and Fuel Systems. Coal Technology Association. Clearwater, FL. March 9 - 13. pp. 707 - 717.
33. Falcone Miller, S., J.L. Morrison, and A.W. Scaroni. 1996. The Effect of Cofiring Coal-Water Slurry Fuel formulated from Waste Coal Fines with Pulverized Coal on NO_x Emissions. Proc. 21st International Technical Conference on Coal Utilization & Fuel Systems. Coal &

Slurry Technology Association. Clearwater, FL. March 5-8. pp. 499-510.

34. Miller, B.G., S.F. Miller, J.L. Morrison, and A.W. Scaroni. 1997. Cofiring Coal-Water Slurry Fuel with Pulverized Coal as a NO_x Reduction Strategy. . Proc. 14th International Pittsburgh Coal Conference. Taiyuan, Shanxi Province, Peoples Republic of China. Sep 23-27.
35. Ashworth, R.A. and T.M. Sommer. 1997. Economical Use of Coal Water Slurry Fuels Produced from Impounded Coal Fines. Proc. Effects of Coal Quality on Power Plants. Electric Power Research Institute. Kansas May 20-22.
36. Zitron, Z., C. Harrison, and D. Akers. 1998. Granuflow™ -- a Technology for Reducing Moisture and Enhancing Coal Handleability. Proc. 23rd International Technical Conference on Coal Utilization and Fuel Systems. Coal Technology Association. Clearwater, FL. March 9 - 13. pp. 741 - 743.
37. Shirey, G., D. Akers, and C. Maronde. 2003. Production of a Composite Fuel from Coal and Biomass. Proc. 28th International Technical Conference on Coal Utilization and Fuel Systems. Coal Technology Association. Clearwater, FL. March 9 – 13.
38. Akers, D.J. 2002. Biomass/coal Composite Fuel Via the Granuflow™ Process. Proc. 27th International Technical Conference on Coal Utilization & Fuel Systems. Coal Technology Association. Clearwater, FL. March 4 – 7. p. 705.
39. Sung, D.J., X. Shao, and B.K. Parekh. 1998. Dewatering/Reconstitution of Fine Clean Coal Slurry. Proc. 23rd International Technical Conference on Coal Utilization and Fuel Systems. Coal Technology Association. Clearwater, FL. March 9 - 13. pp. 745 - 760.
40. Dooher, J. and H. Lebowitz. 2000. Assessment of the Technical and Economic Feasibility of Coal Sludge Slurries. Proc. 25th International Technical Conference on Coal Utilization and Fuel Systems. Coal Technology Association. Clearwater, FL. March 6 - 9. pp. 761 - 768.
41. Anon. 1998. DOE, Antaeus Energy Project to Recover Waste Coal, Expected to Create Jobs in West Virginia, Pennsylvania. DOE News. U.S. Department of Energy, Federal Energy Technology Center, Pittsburgh, PA. Aug 20.

42. Park, W.R., J.C. Blankenship, J.O. Cook, J.R. Herndon, and J.L. Shumate. 1972. Coal Fatal. Interim Report of Retaining Dam Failure. No. 5 Preparation Plant, Buffalo Mining Company Division of the Pittston Company, Saunders, Logan County, West Virginia. United States Department of the Interior, Bureau of Mines. Mt. Hope, WV.
43. W.A. Wahler & Associates. 1973. Analysis of Coal Refuse Dam Failure: Middle Fork Buffalo Creek, Saunders, West Virginia. United States Department of the Interior, Bureau of Mines. Washington, D.C.
44. Lindsay, C., S. Hoyle, and J. Fredland. 2003. Lessons from the Buffalo Creek Disaster. Joseph A. Holmes Safety Association Bulletin. Jan-Feb. pp. 9-12.
45. Thompson, T.J., T.P. Betoney, R.L. Brock, M.A. Evanto, J.W. Fredland, H.L. Owens, S.L. Sorke, and C.A. Weaver. 2001. Report of Investigation: Surface Impoundment Facility Underground Coal Mine Noninjury Impoundment Failure/Mine Inundation Accident, October 11, 2000: Big Branch refuse Impoundment, I.D. No. 1211KY60035-01. United States Department of Labor Mine Safety and Health Administration. Arlington, VA.
46. Li, S. 1999. Operating Experience of Foster Wheeler Waste-Coal Fired CFB Boilers. Proc. 2nd International Symposium on Clean Coal Technology. Beijing, Peoples Republic of China. Nov 8 – 10.
47. Williams, M.A. 2002. Characterization of Coal Wastes as Boiler Fuel. 27th International Technical Conference on Coal Utilization and Fuel Systems. Coal Technology Association. Clearwater, FL. March 4-7. pp. 943 - 954.
48. Anon. 2003. Coal Overview, 1949 – 2001. Annual Energy Review 2001. US Energy Information Administration. Washington, DC.
49. Anon. 2003. Healy Clean Coal Project. Clean Coal Technology Compendium. National Energy Technology Laboratory, USDOE.
50. Svendsen, R.L., and S.A. LeClerc. 1992. Improvements in Plant Design for a Second Generation Waste Coal Fired Power Plant. Proc. Power-Gen '92. Orlando, FL. Nov 17 – 19.
51. Anon. 1997. Ebensburg Power Company Named Plant of the Year. B&W News. Babcock & Wilcox, a McDermott Company.
52. Anon. 2000. Clean Power and Industrial Redevelopment – the Northampton Generating Plant. Heat Engineering. Foster Wheeler, Inc. Clinton, NJ.

CHAPTER 4: WOODY BIOMASS OPPORTUNITY FUELS

4.1. Introduction

The first fuel of importance in pre-industrial economies, including the USA economy, was wood. Until 1850, wood was the dominant energy source, fueling steam engines, riverboat and railroad boilers, and metal smelting operations such as iron furnaces [1-3]. The lumber industry in Maine utilized steam boilers to reduce dependence upon river water and factories organized according to the principles of textile magnate Francis Lowell began employing steam boilers to provide mechanical energy for looms and other machinery [2].

With the advent of wood pulping technology, however, a higher and better use for wood emerged. Mechanical pulping was invented in Germany in 1844, and brought to the US in 1867. The soda pulping process was invented in England in 1851 and brought to the US in 1855; the sulfite process was invented in the USA and commercialized in 1866-1867. Kraft or sulfate pulping was invented in 1884 in Germany and introduced rapidly throughout the industrialized world [2]. The naval stores industry, with a rich tradition, formed one of the bases for the chemicals industry, and with the advent of dissolving pulp processes in the period 1916 – 1920; chemicals from woody biomass reached over 2 million tonnes [2]. Wood distillation products—e.g., methanol, acetone, acetate—also were significant components of the early chemicals industry. Wood and cotton were the main feedstocks of the synthetic materials industry until the 1930's [4].

During the period 1850 – 1900 energy consumption in the USA and throughout the industrializing world increased dramatically, and coal became the dominant fuel. Petroleum production, initiated in Titusville, PA in 1844, provided a glimpse of future energy demand; however solid fossil fuels used under boilers and in coke ovens dominated energy use in all industrialized economies. By 1950, wood and biomass energy declined both absolutely and relatively; in the USA wood energy consumption decreased to about 1.26 EJ (1.2 x 10^{15} Btu) per year [1]. However during the last half of the 20th century, woody biomass became increasingly important, contributing about 3.5 EJ/year to the US energy supply, along with supporting about 7,000 MW_e of generating capacity. In selected European countries such as Finland and Sweden, wood fuel consumption assumed increasing importance as is shown in Figure 4-1. The overwhelming percentage of biomass used is woody biomass.

The contribution of woody biomass to the Swedish economy is illustrative of the use of this fuel. As is shown in Figure 4-2, the dominant use is in the forest products industry, while secondary uses are

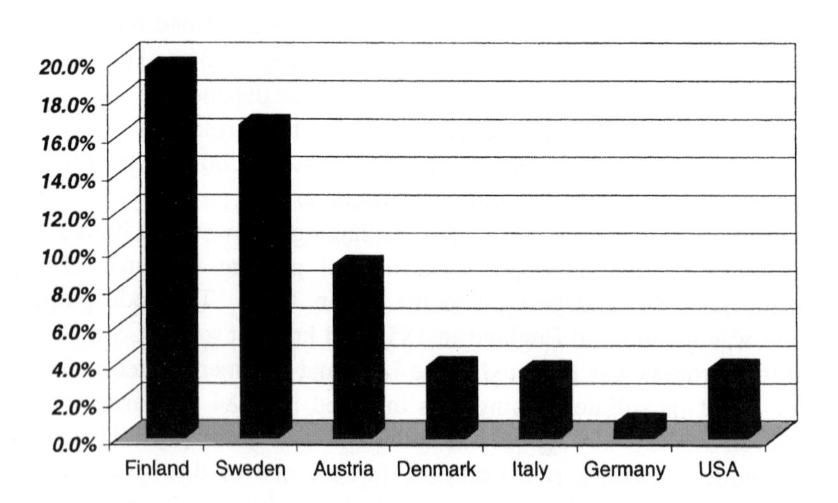

Figure 4.1. Biomass consumption in selected industrialized economies.
Source: [5]

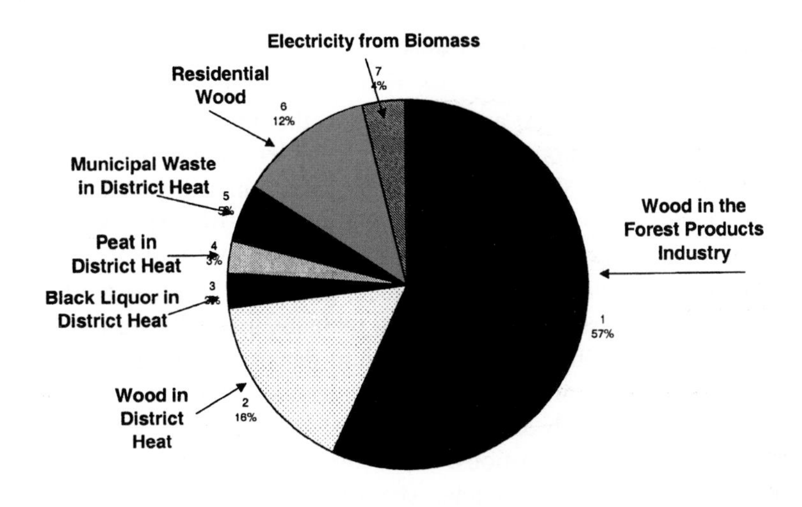

Figure 4-2. Usage of woody biomass in Sweden by economic sector
Source: [5]

for utility applications generating electricity and district heat, and in residential applications.

Wood energy use in the USA followed a similar pattern, with the dominant woody biomass consumption for energy purposes being in the forest products industry. Wood is being consumed in stand-alone electricity generation installations as well—either in independent power producer (IPP) stations or in utility-owned generating installations. In the organization of forest industry activity, typically logs are debarked and either sawn for lumber, peeled for veneer and plywood, or sent to pulp mills. Chips from the sawmill and veneer production processes are sent to pulp mills along with logs, or sent to fiberboard plants. Sawdust and bark is sent to fuel applications—either as hog fuel to boilers or as feedstock to charcoal plants. While most logging residue is left in the forest, some logging residue has been reclaimed as "barky chips" either for pulping or for fuel applications.

4.1.1. Types of Woody Biomass Fuels

Today, in the USA and throughout the industrialized world, several types of woody biomass fuels are used under boilers, in kilns, and in cofiring applications. Cofiring of woody biomass in utility boilers has, in reality, become a new and potentially large market for these opportunity fuels.

Wood fuels available today include, primarily, residues from the wood processing industry. This industry includes sawmills, veneer and plywood mills, particleboard plants, medium density fiberboard (MDF) mills, and pulp mills of various types. It also includes secondary processing industries such as furniture, flooring, cabinetry, and related product manufacturing operations. The residues from the processing facilities are bark, sawdust, planer shavings, sander dust, chips, and mixtures of the above known as "hog fuel". The name "hog fuel" refers to the fact that much of this material is processed through hammermills, also known as "hammer hogs" or "hogs". Spent pulping liquor from pulp and paper mills—including black liquor from kraft mills and red liquor from sulfite mills—provides a special type of wood waste burned only within the pulp and paper community in special chemical recovery boilers designed to return sodium and sulfur to the kraft process as well as providing steam to back pressure turbines and process needs.

In addition to the mill residues identified above, forestry produces significant in-forest residues. In some cases these have been recovered for fuel applications, although such cases are rare due to the relatively high cost of such fuels. The 800 MW wood-to-electricity industry in California during the late 1980's and 1990's was based partly upon recovery of such materials. It is frequently proposed that trees be grown specifically as an energy crop, and such proposals have been made for at least the past 30 years [6 – 12]. Environmental benefits of such short rotation woody crop (SRWC) concepts are cited to promote such concepts. To date this source of wood fuel remains a potential source in the USA and other industrialized economies due to the high costs of such materials.

The final class of woody biofuels available is urban wood waste. Urban wood waste has long been viewed as a potential source of fiber, and recycling programs have made this source potentially promising. Urban wood wastes include manufacturing residues (e.g., residues from the production of manufactured homes or recreational vehicles); construction, demolition, and land clearing (CDL) materials; wood from pallet processors; pallets and other dunnage from the commercial transportation industry; and related "clean" or untreated woody materials. Treated wood wastes also are included in the urban wood waste category. These include used railroad ties, used utility poles, lumber treated with copper-chromium-arsenate (CCA) or pentachlorophenol (PCP) for outdoor applications, and related products. Treated wood waste is a special class of materials, distinct and separate from "clean" or untreated wood waste.

4.1.2. Focus of this Chapter

Woody biomass—particularly mill or forest residue material—can be considered as premium biomass fuel. It is generally sulfur free, low in fuel nitrogen, and low in ash. The clean urban wood waste, while also essentially sulfur free, can be high in fuel nitrogen and somewhat higher in fuel ash than mill or forest residues. Both of these sources of woody biomass opportunity fuels are considered further in this chapter. Because woody biomass can be used for energy purposes or for other applications ranging from pulp mill feedstock to animal bedding, its availability is subject to local conditions. Further, its price is driven by local conditions and local competition; its price can range from \$0.50/GJ (\$0.50/10^6 Btu) to \$1.50/GJ and above. If the biomass is grown for energy purposes, most estimates show costs in the \$2.50 - \$3.50/GJ range. The entire range of availabilities and prices has been experienced by most researchers and wood waste users. Many projects have experienced significant availability and low costs, however supply and demand for these fuels is a local phenomenon.

With increasing attention to carbon sequestration and fossil CO_2 mitigation, increased focus has been given to woody biomass. It is considered CO_2 neutral, and provides a means for utilities and process industries to reduce greenhouse gas emissions.

Most authors also recognize that woody biomass fuels, while low in sulfur and possibly low in nitrogen and ash, are also modest in calorific value. Further, they can be high in moisture content; and they can contain extraneous materials from harvesting or processing. With few exceptions, these fuels are highly reactive with significant volatility.

The generalizations presented above are just that—generalizations. In reality the woody biomass fuels are complex, and highly variable depending upon sources of the wood (including species), types of processing facilities (e.g., sawmills, plywood mills), types of processing (e.g., among sawmills whether the mill is based upon a band saw headrig, a circular saw headrig, or a chipping headrig), storage processes, and the availability of alternative markets for various residues.

This chapter will examine characteristics of woody biomass with attention to sawdust and urban wood waste. It will present specific physical and chemical fuel characteristics for the woody biomass fuels and then consider them through case studies. The case studies will focus upon cofiring woody biomass fuels into coal-fired boilers, since this is the most significant new market for these opportunity fuels.

4.2. Physical and Chemical Characteristics of Woody Biomass Fuels

Woody biomass, either gymnosperms (softwoods) or angiosperms (hardwoods) is inherently anisotropic and hygroscopic; it is porous material, with the porosity caused by the hollow fibers that make up the woody material. Any number of wood technology texts (see, for example, Haygreen and Bowyer [13]) describe the physical structure of woody biomass. The porosity or void volume of wood exists as macropores—tracheids, rays, resin canals and related structures useful in moving moisture and nutrients within the tree. Chemically, woody biomass is comprised of cellulose, the hemicelluloses, one or another type of lignin, and extractives such as pinoresinol, catechin, pinosylvin, and related compounds.

4.2.1. Physical Characteristics of Woody Biomass

On an oven dry weight and volume basis, solid wood typically has a specific gravity of 0.4 – 0.7, with some species being slightly lower and others being slightly higher [14]. Table 1 provides a short table of specific gravities for representative species.

Table 4.1. Specific Gravities for Selected Species (Dry Basis)

Species	Specific Gravity	Species	Specific Gravity
Douglas-fir	0.45 – 0.5	American Elm	0.46 – 0.50
W. Hemlock	0.42 – 0.45	Shagbark Hickory	0.64 – 0.72
Loblolly Pine	0.47 – 0.51	Black Maple	0.52 – 0.57
Pitch Pine	0.47 – 0.52	White Oak	0.60 – 0.68
White Pine	0.34 – 0.35	Willow	0.56 – 0.69

Source: [14]

Specific gravities of fast growing woody biomass species grown under intensive silvicultural regimes are typically lower than specific gravities of naturally grown trees. Short rotation woody crops (SRWC) can have specific gravities in the 0.28 – 0.33 range depending upon species and the age at harvest.

Specific gravity is an indicator of porosity or void volume, since the specific gravity of the solid wood cell wall can be measured at 1.46 – 1.54 depending upon measurement technique; or approximated at 1.50 [15]. On a dry basis, the porosity, or fractional void volume, can be expressed as follows:

$$FVV = (1 - S_{o,o})/1.50 \qquad [4\text{-}1]$$

Where FVV is fractional void volume (porosity) and $S_{o,o}$ is specific gravity on an oven dry weight, oven dry volume basis. Typical fractional void volumes are on the order of 60 – 70 percent.

Moisture content in woody biomass fuels is both a function of the living and growing process, and a function of the manufacturing processes imposed upon a given piece of wood. Since many debarkers and saws are water-cooled devices, much of the moisture in sawdust can be attributed to the processing activities. Heartwood in the hardwoods is typically 32 – 48 percent before processing, hardwood sapwood is typically 31 – 53 percent depending upon individual species [13]. Softwoods have typical heartwood moisture contents of 25 – 49 percent, and sapwood moisture contents of 52 – 71 percent depending upon species [13]. As a practical matter, sawdust or hog fuel received for use under boilers is typically 40 – 50 percent moisture [15, 16].

The combination of specific gravity and moisture content leads to considerations of bulk density—critical in calculating conveying system requirements. For sawdust and hog fuel with moisture contents in the 35 – 45 percent range, typical bulk density measurements have been 0.205 – 0.256 kg/m^3 (16 – 20 lb/ft^3). For woody biomass fuels with lower moisture contents (e.g., 12 percent moisture) such as planer shavings or pallet wastes, bulk densities can be on the order of 0.128 – 0.154 kg/m^3 (10 – 12 lb/ft^3) [16-17].

4.2.2. Proximate and Ultimate Analysis of Woody Biomass

Chemically, wood is composed of cellulose ($C_6H_{10}O_5$); the hemicelluloses including xylan, galactoglucomannans, arabino-glucuronoxylan, arabinogalactan and others in softwoods and glucuronoxylan and glucomannan and others in hardwoods; lignins ($C_9H_{10}O_3(OCH_3)_{0.9\text{-}1.7}$); and extractives such as aliphatic compounds (fats and waxes), terpenes and terpenoids, and phenolic compounds [2, 18]. Additionally woody biomass includes minor quantities of proteins, and inorganic matter. On an extractive-free basis, softwoods typically contain 45 – 50 percent cellulose, 25 – 35 percent hemicelluloses, and 25 – 35 percent lignin while hardwoods contain about 40 – 55 percent cellulose, 24 – 40 percent hemicelluloses, and 18 – 25 percent lignin. Extractives typically comprise 1 – 5 percent of the wood [15, 18]. Table 4-2 presents extractive-free analyses for three wood types.

Table 4-2. Composition of Three Wood Species (values in weight %)

Component	Wood Species		
	Spruce	Pine	Birch
Cellulose	43	44	40
Hemicelluloses	27	26	39
Lignin	28.6	29	21
Extractives	1.8	5.3	3.1
Protein	1.3	1.2	2.5
Inorganic Matter	0.4	0.4	0.3

Source: [19]

Given the chemical compositions of these various components of wood, the higher heating value of woody biomass can be approximated by the following formulae:

$$HHV\ (\text{`MJ/kg}) \cong HC \times 17.52 + (1\text{-}HC) \times 26.72 \qquad [4\text{-}2]$$

Where HC is the holocellulose fraction of wood (cellulose and the hemicelluloses) and 1-HC represents the lignin and extractives fraction [2]. Alternatively this equation is as follows:

$$HHV(\text{Btu/lb}) \cong HC \times 7527 + (1\text{-}HC) \times 11{,}479 \qquad [4\text{-}3]$$

From a fuels perspective, chemical composition is typically presented as proximate and ultimate analysis, shown in Table 4-3 for selected wood species. Note the differences between the virgin wood fuels and the urban wood waste. That difference goes beyond lower moisture content and higher ash content, and includes a very high concentration of fuel nitrogen. The fuel nitrogen comes from the urea formaldehyde and related glues used to manufacture plywood, particleboard, and paneling. On a kg/GJ (lb/10^6 Btu) basis, these fuels have the following fuel nitrogen concentrations: pine, 0.103 (0.24); red oak, 0.267 (0.62); mixed hardwood/softwood sawdust, 0.206 (0.48); and urban wood waste, 0.72 (1.67).

4.2.3. Reactivity of Woody Biomass

Reactivity of woody biomass can be approximated by the data in Table 4-3; alternatively it can be measured more precisely and amplified by the use of drop tube reactor (DTR), Thermogravimetric analysis (TGA), and Carbon 13 Nuclear Magnetic Resonance (^{13}C NMR) techniques. These techniques, discussed in some detail in Chapter 2, have been used to characterize woody biomass. The results complement the data in Table 4-3.

4.2.3.1 Drop tube reactor measurements.

DTR experiments conducted by The Energy Institute of Pennsylvania State University have been used to characterize the reactivity of both sawdust and urban wood waste. The sawdust tested in the DTR has been shown in Table 4-3. The urban wood waste was synthetically constructed from plywood, particleboard, and paneling to achieve a material comparable to the urban wood waste burned at the

Table 4-3. Typical Fuel Characteristics of Selected Wood Fuels

Analysis	Fuel Type			
	Pine	Red Oak	Mixed Sawdust	Urban Wood Waste
Moisture (%)	45.0	28.8	40.0	30.8
Proximate Analysis (dry wt %)				
Fixed Carbon	15.2	19.0	19.0	18.1
Volatiles	84.7	79.5	80.0	76.0
Ash	0.1	1.5	1.0	5.9
Total	100.0	100.0	100.0	100.0
Ultimate Analysis (dry wt %)				
Carbon	49.1	51.6	49.2	48.0
Hydrogen	6.4	5.8	6.0	5.5
Oxygen	44.0	40.6	43.0	39.1
Nitrogen	0.2	0.5	0.4	1.4
Sulfur	0.2	<0.1	<0.1	0.1
Ash	0.1	1.5	1.0	5.9
Total	100.0	100.0	100.0	100.0
Higher Heating Value (dry basis)				
MJ/kg	19.79	18.78	19.56	19.47
Btu/lb	8502	8069	8400	8364

Sources: [20 - 22]

Bailly Generating Station (see Tillman, [17, 23]. This synthetic urban wood waste is shown in Table 4-4. Note the high concentration of fuel nitrogen. Such concentrations of nitrogen in urban wood waste are routinely measured when evaluating such fuels. The nitrogen is contained in the wood and, in glues and coatings employed in products.

DTR experiments were carried out in an argon atmosphere at temperatures from 400°C to 1700°C (750°F to 3090°F) in order to determine the devolatilization reactivity of both sawdust and urban wood waste. Using the equations shown in Chapter 2, Arrhenius equations were then determined. It should be noted that these are reactivity measurements based upon bulk furnace temperatures rather than particle temperatures. These measurements provide the data necessary for computational fluid dynamics (CFD) modeling.

Table 4-4. Fuel Analysis of Components of Urban Wood Waste Composite

Analysis	Fuel Type			
	Paneling	Particleboard	Plywood	Composite
Proximate Analysis (dry wt %)				
Fixed Carbon	18.95	19.85	17.51	18.34
Volatiles	79.93	79.74	81.61	80.93
Ash	1.12	0.41	0.88	0.73
Total	100.0	100.0	100.0	100.0
Ultimate Analysis (dry wt %)				
Carbon	50.47	51.00	52.2	51.44
Hydrogen	5.83	6.34	6.27	5.97
Oxygen	39.59	39.49	40.35	40.36
Nitrogen	2.97	2.72	0.29	1.45
Sulfur	0.02	0.03	0.01	0.04
Ash	1.12	0.41	0.88	0.73
Total	100.0	100.0	100.0	100.0
Higher Heating Value (*)				
MJ/kg	18.62	19.54	19.69	19.01
Btu/lb	8000	8395	8457	8166

Note: heating values are calculated. Sources: [20, 24]

Figures 4-1 through 4-3 show the devolatilization or pyrolysis reactivity profiles of sawdust and urban wood waste; note that there is a two-stage devolatilization for the sawdust; this is not true for the urban wood waste where the material has been subjected to drying and weathering. The low temperature devolatilization is largely completed before the urban wood becomes a fuel. The consequence of this phenomenon is that the fresh sawdust has an even greater potential for NO_x reduction than the urban wood waste.

4.2.3.2. Char oxidation kinetics.

Char oxidation kinetics were determined at The Energy Institute of Pennsylvania State University for sawdust, using the TGA technique. This technique is described in Chapter 2. The results of this analysis are shown in Figure 4-4. The char utilized in these experiments was obtained

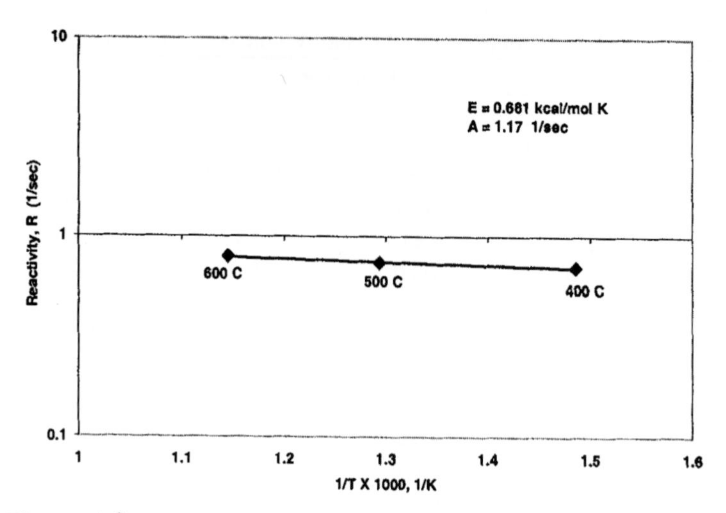

Figure 4-3. Low Temperature Devolatilization Reactivity of Sawdust
Source: [22]

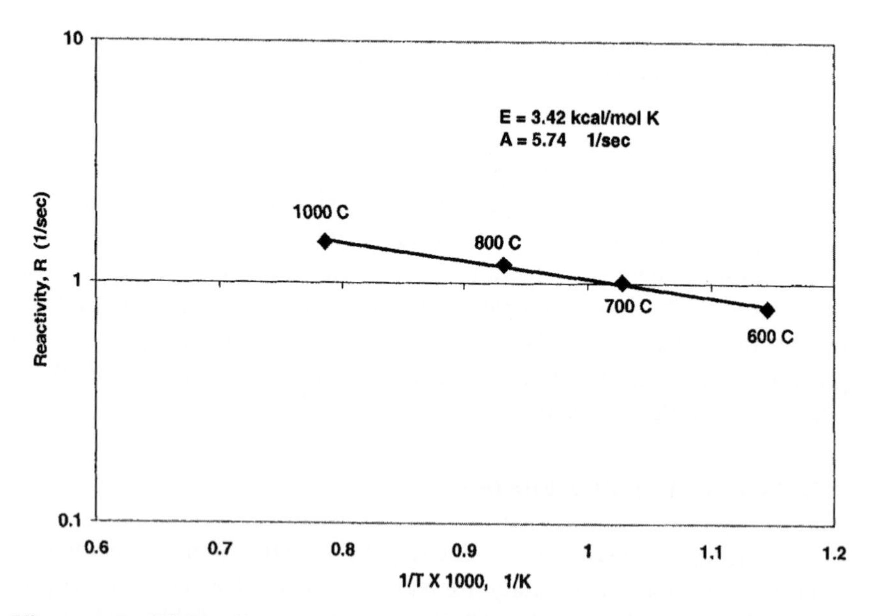

Figure 4-4. Higher Temperature Devolatilization Reactivity of Sawdust
Source: [22]

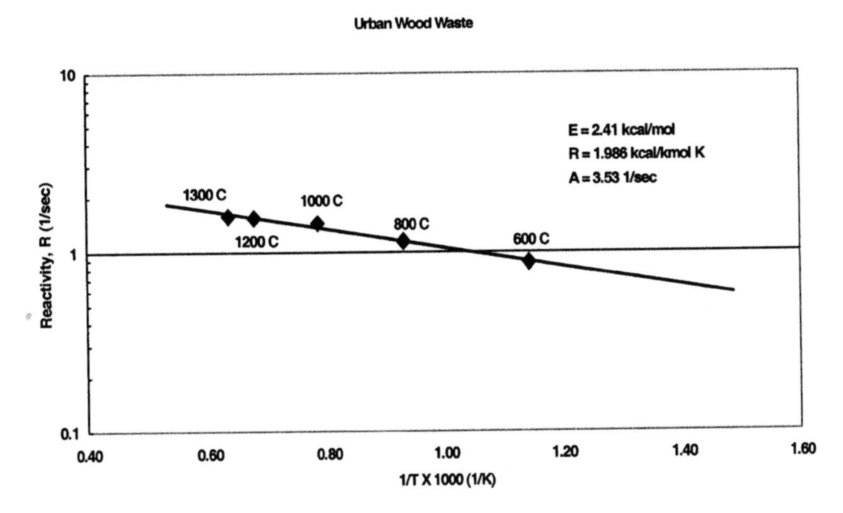

Figure 4-5. Devolatilization Reactivity of Urban Wood Waste
Source: [20]

from the devolatilization of sawdust at 1700°C (3090°F), in order to use a sample characteristic of a high temperature combustion environment. While this temperature used to generate the char is somewhat higher than that experienced in typical pulverized coal firing, it is somewhat lower than that experienced in cyclone firing and is therefore reasonably representative of modern coal-fired boiler combustion.

Table 4-5 summarizes the reactivity determination for the sawdust and urban wood waste, and compares them to Black Thunder, a Powder River Basin (PRB) subbituminous coal and to a Pittsburgh #8 bituminous coal. Note the extreme reactivity of the woody biomass compared even to the relatively reactive PRB coal.
Another direct measure of reactivity is the determination of maximum volatile yield from pyrolysis or devolatilization of any fuel. Figure 4-5 compares the maximum volatile yield of urban wood waste and sawdust to the maximum volatile yield of Black Thunder and Pittsburgh #8 coal. Note that the maximum volatile yield for both woody biomass samples exceeded 90 percent. This is in direct contrast to the coals, and to the petroleum coke as shown in Chapter 2.

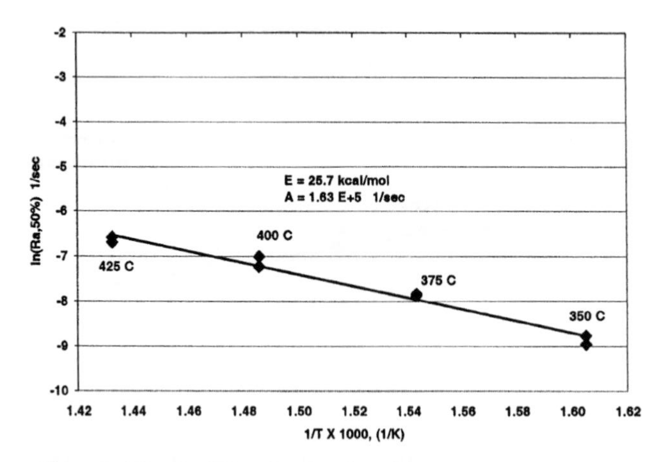

Figure 4-6. Char Oxidation Kinetics for Sawdust.
Source: [23]

Table 4-5. Short Table of Kinetic Parameters for Woody Biomass Devolatilization and Char Oxidation Compared to Reference Coals[1]

Fuel	Pre-exponential constant A (1/sec)	Activation energy E (kJ/mol)
Devolatilization		
Sawdust (400 – 600°C)	1.17	0.681
Sawdust (600 – 1000°C)	5.74	3.42
Urban Wood Waste (400 – 1600°C)	3.53	2.41
Black Thunder PRB	59.1	9.53
Pittsburgh #8 Bituminous	66.2	10.3
Char Oxidation		
Woody Biomass	1.63×10^5	25.7
Black Thunder PRB	7.61×10^4	27.4
Pittsburgh #8 Bituminous	3.72×10^5	32.3

Note: [1]The kinetics presented here are based upon bulk DTR temperatures, rather than particle temperatures. They have to be adjusted appropriately to be used in many CFD programs.
Sources: [22, 24 - 25]

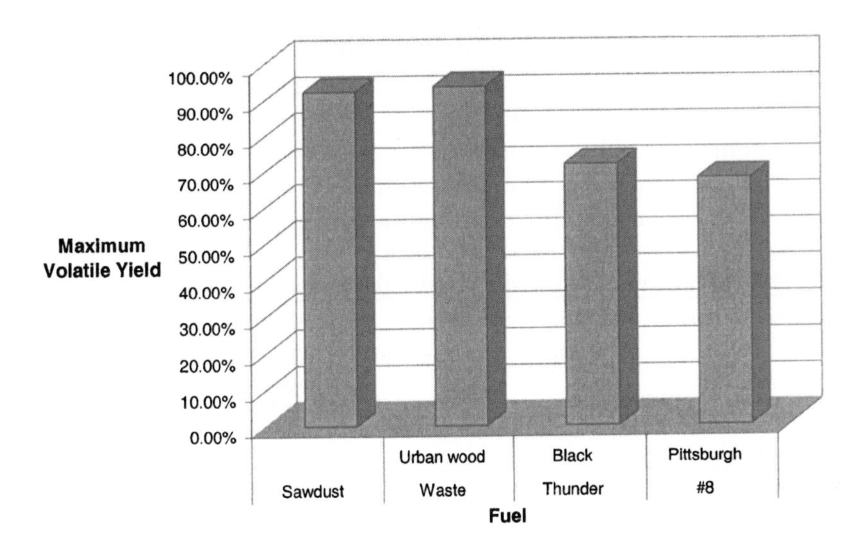

Figure 4-5. Maximum Volatile Yield for Sawdust, Urban Wood Waste, Black Thunder, and Pittsburgh #8 Coal
Source: [24].

4.2.3.3. Carbon 13 nuclear magnetic resonance.

The reactivity shown in the above figures and table relates directly to structure. ^{13}C NMR experimentation at The Energy Institute of Pennsylvania State University was used to determine the structural characteristics of woody biomass relative to various coal and opportunity fuel samples. These experiments demonstrated that woody biomass is less aromatic than any of the coals. Further the number of carbon atoms per aromatic cluster is lower in woody biomass than in any other fossil or opportunity fuel, and the percentage of carbon atoms existing in methoxy (OCH_3) functional groups is higher in woody biomass than in any of the coals or opportunity fuels [23 - 24].

The overall reactivity picture, integrating the proximate and ultimate analysis data along with DTR experimentation and ^{13}C NMR testing, is shown in Table 4-6.

The data in Table 4-6 highlight the interrelationships among analytical techniques. Note the ^{13}C NMR measures of volatility including aromaticity, number of aromatic carbons per cluster, and percentage of carbon atoms in methoxy functional groups corresponds well to the volatile/fixed carbon ratio and the maximum volatile yield for any fuel. Further, it corresponds to the activation energies measured in the DTR.

Table 4-6. Measures of Reactivity for Woody Biomass Fuels

Reactivity Measure	Urban Wood Waste	Fresh Sawdust	Black Thunder PRB	Pittsburgh #8 Bituminous Coal
Volatile/Fixed Carbon Ratio	4.20	4.21	0.92	0.61
H/C Atomic Ratio	1.38	1.46	0.83	0.77
O/C Atomic Ratio	0.61	0.66	0.21	0.05
Maximum Volatile Yield (%)	97.2	92.7	70.0	52.8
Aromaticity (%)	8	8	57	70
Average Number of Aromatic Carbon Atoms per Cluster	6	6	10	15
Concentration of Carbon Atoms in Methoxy Functional Groups	---	8.62	4.21	0
Devolatilization Activation Energy (kcal/mol)	2.41	0.681, 3.42 (*)	9.53	10.3
Char Oxidation Activation Energy (kcal/mol)	25.7(**)	25.7	27.4	32.3

(*) Sawdust and has 2-step devolatilization mechanisms, with initial devolatilization having very low activation energies, and with subsequent devolatilization being volatile stripping from the char matrix, with higher activation energies.

(**) Char oxidation activation energy considered same for all woody biomass

Sources: [22 - 25]

4.2.3.4. Nitrogen reactivity.

The high reactivity of woody biomass supports the use of this fuel as a means for reducing NO_x emissions. Further, the fuel nitrogen in woody biomass is also highly reactive. The experiments conducted by Baxter and co-workers [26] and previously referenced in Chapter 2 document the utility of comparing the rate of nitrogen evolution during devolatilization or pyrolysis with the overall rate of volatile evolution from any fuel. When volatile nitrogen evolution lags behind total volatile evolution—which is typically almost identical to volatile carbon evolution—then the control of NO_x formation by staged combustion is more difficult. Conversely, if volatile nitrogen evolution occurs more rapidly than total volatile evolution—or volatile carbon evolution—then staged combustion can be more effective.

Figures 4-6 through 4-8 depict volatile nitrogen evolution from woody biomass. Figure 4-6 shows volatile nitrogen evolution from sawdust compared to volatile carbon evolution while Figure 4-7 depicts volatile nitrogen evolution from sawdust normalized against total volatile matter evolution. Figure 4-8 compares volatile nitrogen evolution to volatile carbon evolution for the urban wood waste sample prepared by The Energy Institute of Pennsylvania State University.

The data in Figures 4-6 and 4-8 illustrate the clear differences between sawdust and urban wood waste. With fresh sawdust, the nitrogen evolves very rapidly—far more rapidly than the evolution of volatile carbon or the evolution of total volatile matter. The volatile nitrogen evolution for urban wood waste essentially mirrors that of the volatile carbon, with a slight lag in the initial stages of volatile evolution. Sawdust is clearly a more volatile fuel, and a better material for combustion if NO_x control through fuel blending is sought [24].

The volatile nitrogen evolution for biomass can be compared to that for bituminous or PRB coal through the calculation of nitrogen/carbon (N/C) atomic ratios in char resulting from devolatilization at any given temperature. The results for any given fuel are then normalized to the initial N/C atomic ratios of the incoming fuel.

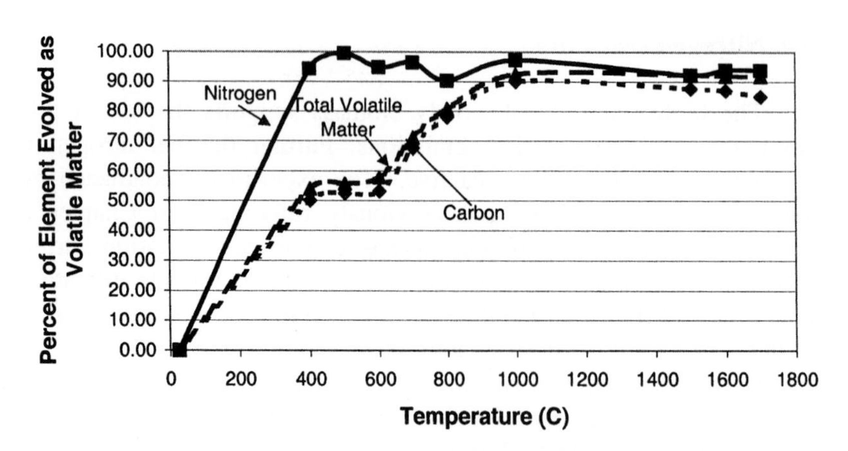

Figure 4-6. Volatile nitrogen, volatile carbon, and total volatile matter evolution from sawdust as a function of temperature
Source: [23]

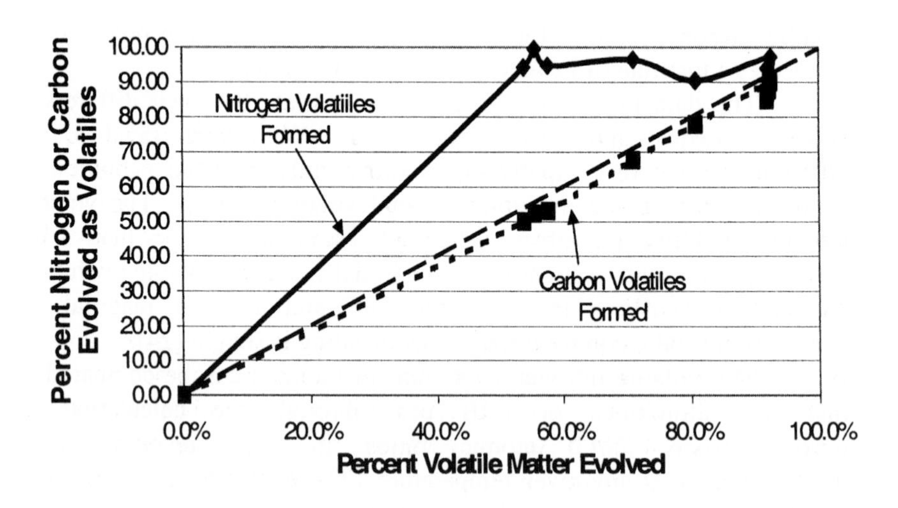

Figure 4-7. Volatile nitrogen and carbon evolution from sawdust normalized to total volatile matter evolution
Source: [23]

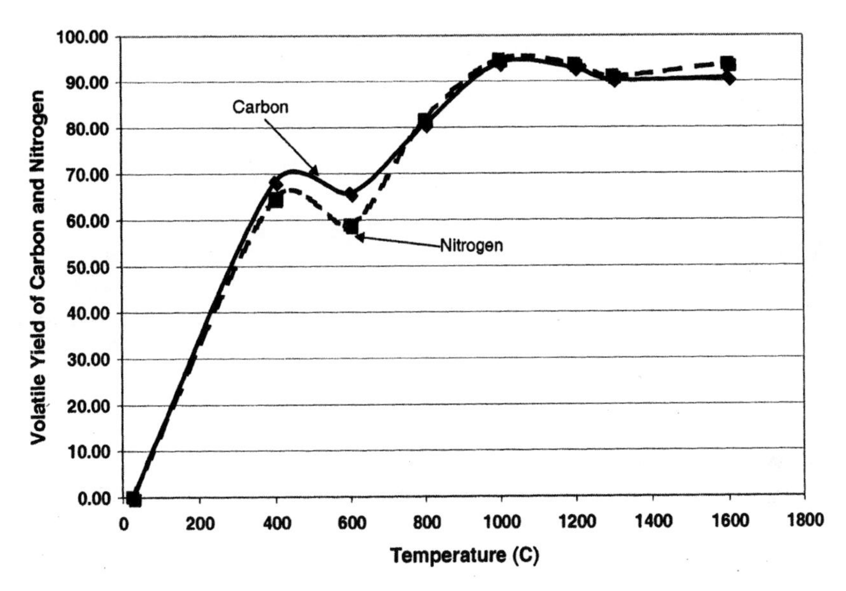

Figure 4-8. Volatile nitrogen and carbon evolution from urban wood waste as a function of temperature
Source: [23]

In these calculations, if the N/C ratio is declining, then the volatile nitrogen is evolving more rapidly than the volatile carbon. If the N/C ratio is increasing, then the volatile carbon is evolving more rapidly than the volatile nitrogen. Figure 4-9 presents these results for sawdust, urban wood waste, Pittsburgh #8 bituminous coal, and Black Thunder PRB subbituminous coal.

The clear implication from the data in Figure 4-9 is that sawdust is the preferred woody biomass fuel in terms of combustion while minimizing NO_x formation. Urban wood waste exhibits a slight lag between the evolution of volatile nitrogen and volatile carbon in the low temperature (e.g., <600°C) region, and then performs favorably with respect to NO_x management. Urban wood waste is somewhat less favorable than Black Thunder as a representative PRB coal; it is more favorable than the Pittsburgh #8 bituminous coal.

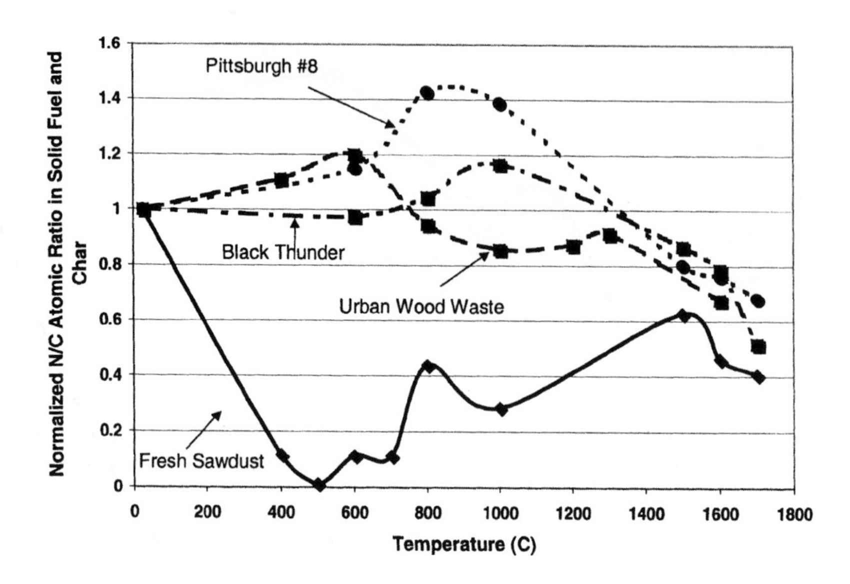

Figure 4-9. Atomic nitrogen/carbon ratios for char from fresh sawdust, urban wood waste, Black Thunder, and Pittsburgh #8 coal normalized to the N/C ratios for the incoming fuel as a function of pyrolysis temperature

The nitrogen evolution data, like the overall reactivity data, can be related to a 2-stage model where initial volatiles and gases are driven off at low temperatures—temperatures that can be encountered in wood drying systems (e.g., lumber dry kilns, veneer dryers). These volatiles result from extractives, some hemicelluloses, and functional groups such as amines. Further, cellulose and the hemicelluloses may be converted to "active cellulose" when being heated at low temperatures [27, 28].

The woody biomass remaining, and devolatilized at higher temperatures—when bulk furnace temperatures exceed 600°C and particle temperatures are considerably higher—may be more tightly bound and may include a higher portion of lignins. Its pyrolysis requires higher activation energies. The nitrogen in sawdust is largely in amine structures (e.g., NH, NH_2, NH_3) that readily evolve from the solid fuel matrix. The nitrogen in urban wood waste is largely in the form of glues such as urea-formaldehyde; its evolution as volatile nitrogen occurs less rapidly.

The total volatile evolution data, and nitrogen volatile evolution data clearly support the conclusions by Harding and Adams [29 - 30] that sawdust can be as effective a NO_x control fuel—through the reburn process—as natural gas or coal. Research at the University of Utah and Reaction Engineering International clearly documented the applicability of woody biomass as a reburn fuel, capitalizing upon its reactivity.

The volatility data associated with sawdust can be considered representative of all virgin woody biomass—whole tree chips and short rotation woody crops. However the reactivity of woody fuels, as a practical matter, is significantly influenced by the particle size of the material being fired. Junge [31 - 32] clearly demonstrated this phenomenon; the cofiring research at the Allen Fossil Plant [23, 33] again demonstrated this influence and its relationship to NO_x management.

4.2.4. Inorganic Matter in Woody Biomass

As shown previously in Table 4-4, virgin woody biomass is a low ash or low inorganic matter fuel. Inorganic matter, expressed as a percentage of the total fuel mass or in g/kJ (lb/10^6 Btu) is routinely low for wood fuels, particularly when compared to coals. Urban wood waste is much higher in ash content as shown in Table 4-4; the difference is in the presence of dirt in the urban wood waste, along with wood coatings (e.g., paints) and inorganic matter in many of the glues used for producing particleboard, plywood, paneling, furniture, and other products that end up in the urban wood waste pile.

Tables 4-7 and 4-8 present bulk ash chemistries of various wood fuels obtained during test programs. Table 4-7 focuses upon virgin wood material, including sawdust and clean tree trimmings from the urban forest. Table 4-8 presents representative data from some urban wood wastes plus short rotation woody crops. Wide variations in ash compositions can be observed between the various woody fuel samples.

Table 4-7. Ash Elemental Analysis of Typical Wood Fuels

Analysis	Fuel Type			
	Pine Wood	Hardwood Sawdust	Pine Bark	Mixed Hardwood, Softwood
Elemental Composition (wt %)				
SiO_2	57.2	23.7	0.40	23.5
Al_2O_3	13.4	4.10	0.30	5.10
TiO_2	1.16	0.36	0.00	0.10
Fe_2O_3	5.94	1.65	0.20	2.10
CaO	8.75	39.9	40.6	33.6
MgO	3.35	4.84	5.10	5.10
Na_2O	1.38	2.25	0.30	0.20
K_2O	4.94	9.81	26.5	12.0
P_2O_5	1.44	2.06	11.5	4.80
SO_3	0.05	1.86	3.0	1.60
Reducing Ash Fusion Temperatures (°C)				
Initial Def.	---	1246	1375	1230
Spherical	---	1414	1504	1240
Hemispherical	---	1417	1506	1245
Fluid	---	1424	1507	1290
Oxidizing Ash Fusion Temperatures (°C)				
Initial Def.	---	1397	1340	1210
Spherical	---	1406	1525	1225
Hemispherical	---	1408	1650	1250
Fluid	---	1414	1650	1275

Sources: [16-17, 20]

To a large extent, the SiO_2 content reflects extraneous dirt in the sample—above about 5 – 10 percent silica in the inorganic constituents. Note the wide variation in CaO and Al_2O_3 among the samples shown. Note, also, the high concentration of K_2O in samples shown. It can be readily observed that the woody biomass ash is almost always high in CaO and K_2O; the Na_2O can be relatively low although wood fuel derived from salt water-stored logs can have significant sodium concentrations. Alternatively, some soils contribute Na_2O to the inorganic fraction.

Table 4-8. Ash Elemental Analysis of Typical Wood Fuels

Analysis	Fuel Type			
	Shredded railroad ties (8.58% ash, o.d. basis) *	Hybrid poplar from mine reclamation (9.8% ash, o.d. basis	Willow grown as short rotation woody crop (1.82% ash, o.d. basis)	Green sawdust, east Tennessee (1.06% ash, o.d. basis)
Elemental Composition (wt %)				
SiO_2	47.26	59.21	4.39	21.30
Al_2O_3	16.82	16.69	1.52	4.55
TiO_2	1.25	0.67	1.35	0.61
Fe_2O_3	12.34	8.20	1.02	2.01
CaO	9.18	4.90	54.09	39.83
MgO	4.21	2.05	4.18	6.61
Na_2O	2.73	0.31	0.75	2.20
K_2O	2.15	8.10	12.50	9.48
P_2O_5	0.64	0.62	1.36	2.99
SO_3	0.65	0.09	3.57	1.86
Reducing Ash Fusion Temperatures (°C)				
Initial Def.	1108	---	---	1392
Spherical	1120	---	---	1397
Hemispherical	1124	---	---	1399
Fluid	1136	---	---	1403
Oxidizing Ash Fusion Temperatures (°C)				
Initial Def.	1170	---	---	1356
Spherical	1187	---	---	1386
Hemispherical	1194	---	---	1388
Fluid	1207	---	---	1390

Note: % ash on an o.d. basis refers to the percent ash in the total fuel on an oven dry basis. Source: [16]

4.2.4.1. Slagging and fouling potential.

Slagging and fouling potential for woody biomass, fired exclusively or with coal, has been extensively studied both in the US and

Europe [34 – 39]. Estimates of the slagging and fouling potential can be obtained by determining the concentration of alkali metals in the fuel, expressed in kg/GJ (lb/10^6 Btu). Miles et. al. [40] developed a first approximation technique whereby fuels with <0.17 kg/GJ (0.4 lb/10^6 Btu) Na_2O and K_2O are considered to have minimum slagging and fouling potential, fuels with 0.17 – 0.34 kg/GJ (0.4 – 0.8 lb/10^6 Btu) are considered to be possible slagging fuels or fuels with moderate slagging potential, and fuels with >0.34 kg/GJ are considered to have certain slagging potential. Table 4-9 considers several woody biomass fuels.

Note that the virgin woody biomass fuels naturally grown, or grown for wood products such as lumber and pulp and paper, typically have low slagging and fouling potential; the short rotation woody crop materials and the urban wood wastes have moderate slagging potential. Very few wood fuel samples exhibit high slagging potential.

4.2.4.2. Ash reactivity and deposition potential.

Baxter [41 - 42] and co-workers have shown that ash constituents alone do not govern the behavior of inorganic matter in a fuel; rather, the

Table 4-9. Estimated Slagging and Fouling Potential of Various Wood Fuels

	Woody Biomass Fuel				
	Pine Chips	White Oak	Hybrid Poplar	Urban Wood Waste	East Tennessee Sawdust
Heat Content (dry basis)					
MJ/kg	19.88	18.99	19.02	19.01	19.10
Btu/lb	8550	8165	8178	8174	8213
Ash %	0.7	0.4	1.9	6.0	0.65
Total Alkali in Ash (K + Na)					
% in Ash	3.0	31.8	19.8	6.2	15.49
kg/tonne	0.2	1.16	3.77	3.72	1.01
lb/ton	0.4	2.3	7.5	7.4	2.01
kg/GJ	0.03	0.06	0.20	0.20	0.05
lb/10^6 Btu	0.07	0.14	0.46	0.46	0.12

Sources: [16, 40]

reactivity of those constituents along with their total concentration governs reactivity. Reactivity is typically is measured by chemical fractionation of the ash involving successively leaching fuel samples in water, ammonium acetate, and hydrochloric acid [34, 43]. In this procedure, the leachate and the residues are successively measured to complete mass balances for each element. From this procedure, the percentage of any element (e.g., Ca, K) found in the water leachate, the ammonium acetate leachate, the hydrochloric acid leachate, or in the residue is measured [34, 43]. The material found in the water and ammonium acetate leachate is most reactive; the material found in the residue is largely unreactive—bound up in clays and other non-reactive material [34, 43].

Table 4-10 provides chemical fractionation data for woody biomass, developed by Miller and Miller [34]. Note that the chemical fractionation tests were performed on a typical virgin wood sample.

These data show very high chemical reactivity for the alkali metals and alkali earth metals in the wood fuel. Note, particularly, the very high availability and reactivity of the calcium in wood fuel. This is highly significant given the calcium concentrations in woody biomass ash as shown previously in Tables 4-7 and 4-8. Because of this reactivity there is significant potential for the alkali and alkaline earth metals to react with sulfur and (if present) chlorine in coal being burned with woody biomass. This has some concern regarding the potential of reactive biomass ash to cause waterwall wastage from the formation of

Tab le 4-10. Chemical Fractionation of Pine Shavings Ash (values in wt %)

Element	Fraction		
	Water Soluble and Ion Exchangeable	Acid Soluble	Insoluble
Potassium	65	0	35
Sodium	77	0	23
Calcium	92	6	2
Magnesium	72	10	18
Aluminum	30	0	70
Silicon	29	0	71

Source: [34]

alkali iron trisulfates, and has some potential to deactivate selective catalytic reduction (SCR) catalysts due to the calcium. Both phenomena merit consideration when evaluating woody biomass as a source of opportunity fuels. The potential is mitigated to a large extent, however, by the very low concentrations of inorganic constituents in the woody biomass. As is shown in Table 4.4, the ash content of woody biomass is typically on the order of 0.5 – 1.5 percent. The total concentration of alkali metals and alkaline earth metals is very low indeed.

The chemical fractionation data shown in Table 4-10 are quite similar to chemical fractionation data developed by Baxter and co-workers [41 - 42] for woody biomass. The alkali metals and alkali earth minerals are readily available and highly reactive in wood fuels. The low slagging factors for these fuels is strictly a function of the low concentrations of ash in the total fuel mass. Unterberger and coworkers [34] have shown similar results as well.

Yan and co-workers [36] have related slagging and fouling, and ash reactivity, to computer controlled scanning electron microscopy (CCSEM). This work, focusing upon particle sizes in the flyash, has shown that the woody biomass tends to produce higher concentrations of fine (1 – 3 μm) particles while reducing the concentration of moderate sized (3 – 10 μm) particles. Korbe et. al. [37] have related chemical fractionation data to CCSEM analyses of ash reactivity for biomass fuels, as have Folkedahl et. al. [39]. This research has clearly shown that wood waste is significantly less reactive, and has less slagging and fouling potential, than wood grown for energy purposes.

4.2.4.3. Trace metal concentrations.

Trace metal concentrations in woody biomass are typically lower than trace metal concentrations in coal [17, 20, 44]. Table 4-11 summarizes trace metal concentrations from wood fuel found at various locations [44 - 47]. Table 4-12 summarizes trace metal concentrations in the urban wood waste cofired at the Bailly Generating Station. It should be noted that the mercury analyses shown in earlier references are typically below detection limits due to the nature of the fuel testing. Analyses performed after about 1998 are based upon the more recently developed gold amalgam technique, where detection limits are in the low ppbw or high pptw.

Testing at the Albright and Willow Island Generating Stations of Allegheny Energy Co., LLC. has shown very low mercury concentrations in sawdust. The sawdust burned at Albright was tested for mercury concentration, and it shows 0.003 – 0.009 mg/kg (dry basis fuel), or 3 – 9 ppbw [45]. This compares to mercury concentrations in Pittsburgh seam coal of 0.18 mg/kg (dry fuel basis) [48]. Even when measured in g/GJ or lb/10^6 Btu, the mercury concentrations in woody fuels are substantially less than those measured for most ranks of coal. This low mercury content will become increasingly important due to regulatory conditions.

Table 4-11. Trace Metal Concentrations in Various Woody Biomass Fuels (mg/kg wood ash)

Metal	Woody Biomass Fuel			
	Big Valley Lumber, Bieber, CA	Burlington Electric, Burlington, VT	Georgia Pacific, Lyons, NY	Hog fuel from a Pulp Mill in Pacific Northwest
As	4.6	6.3		0.475
Ba	130	---	---	51.5
Be	0.1	---	---	BDL
Cd	1.5	16	6.7	BDL
Cr (total)	22	25	16.8	128.4
Cr (+6)	---	---	---	0.063
Cu	40	70	76.9	5.625
Pb	38	70	58.8	2.71
Hg	<0.05	---	---	BDL
Mo	14	3.0	---	---
Ni	11	---	---	137.3
Se	5.0	---	---	---
Ag	<0.08	---	---	---
V	26	27	---	---
Zn	130	560	310	99

Sources: [15, 44 - 48].

Table 4-12. Trace Metal Concentrations in the Urban Wood Waste Burned at Bailly Generating Station

Metal	Concentration (mg/kg)
Arsenic	2.145
Chromium	6.57
Lead	2.922
Mercury	0.0126
Nickel	2.645
Vanadium	3.060

Source: [17]

4.3. Using Woody Biomass as an Opportunity Fuel in Dedicated Boilers

Given the unique characteristics of woody biomass, it has been frequently used as an opportunity fuel. It is a common fuel—the dominant fuel—in the wood products and pulp and paper industries. In such applications it is used in boilers designed specifically to fire wood waste. Large wood-fired boilers have been installed for both forest products industry applications and for selected utility applications.

In industrial settings, wood-fired boilers can be sized anywhere from 3 kg/sec ($25x10^3$ lb/hr) saturated steam to over 88.3 kg/sec ($700x10^3$ lb/hr) superheated steam suitable for power generation. The larger boilers are of concern in this text.

4.3.1. Typical Large Woody Biomass-fired Boiler Configurations

Typical large-scale wood fired boilers used in the pulp and paper industry for both process steam and power generation are on the order of 37.8 kg/sec ($300x10^3$ lb/hr) – 75.7 kg/sec ($600x10^3$ lb/hr) of steam typically produced at 86.2 – 100 bar (1250 – 1450 psig)/510°C (950°F). Such boilers are not reheat boilers; rather, the steam generated from these boilers passes through either a backpressure or non-condensing extraction turbine, and then is exhausted typically at 3.4 bar (50 psig) – 10.3 bar

(150 psig) for process applications. Wood fired boilers of this type are typically grate-fired units, although circulating fluidized bed (CFB) boilers are becoming increasingly popular for power boiler applications. With the advent of high pressure/high temperature CFB units, pulp mills have used them increasingly to overcome problems of variability in the fuel composition and moisture content. Further, several pulp mills have turned more to CFB boilers because they can accept wide variations in the fuel blend. The literature concerning such systems is replete with citations (see, for example, [15, 31 - 32, 44, 49 - 52]).

The typical grate-fired boiler for woody biomass can come in a number of firing system configurations including pile burner, inclined grate, and spreader-stoker designs. These boilers fire a range of solid fuels including bark, hogged wood waste, sawdust, orchard and vineyard prunings, municipal solid waste, refuse-derived fuel (from municipal solid waste), and blends of the above with such opportunity fuels as tire-derived fuel (see Chapter 6). Examples of such large–scale boilers in the utility industry include the McNeill Generating Station in Burlington, VT, the Kettle Falls Generating Station just north of Spokane, WA, the Shasta Generating Station outside of Redding, CA, and various cogeneration power boilers built at large pulp mills of Weyerhaeuser, Georgia Pacific, International Paper, and other companies. Several of these boilers fire principally wood waste, supplemented by coal, oil, natural gas, TDF, or other combustible materials. Many are designed to fire a multitude of fuels in various combinations.

Fuel for woody biomass-fired boilers—either grate fired or CFB fired—is typically sized depending upon fuel type, with specifications for woody biomass being typically 25 mm x 0 mm (1" x 0"). In these systems the maximum firing capacity of the boiler is typically around 630 – 740 GJ/hr or 175 – 206 MW_{th} (600 – 700 x 10^6 Btu/hr) for inclined grate and spreader-stoker systems; pile-burning systems are typically much smaller. The maximum capacity is a function of both materials handling logistics and grate design. Typical combustion parameters for these systems are shown in Table 4-13.

Studies of dedicated boilers have demonstrated that, under certain circumstances, significant deposition of inorganic deposits can occur either as slagging (in the furnace zone) or as fouling (in the convective passes of the boiler). These studies, led by Baxter et. al. [40-41], have shown that the alkalinity of the woody biomass ash—and the reactivity of

Table 4-13. Typical Parameters for Woody Biomass-Fired Boilers

	Boiler Type		
	Pile Burner	Inclined Grate	Spreader-stoker
Maximum fuel moisture (%)	65	60	55
Grate heat release, (kW/m^2)	63 - 79	63 - 126	252 – 315
Grate heat release ($10^3 Btu/ft^2$-hr)	200-250	200-400	800-1000
Furnace heat release, (kW/m^3)	2.8	2.8 – 3.6	4.2 – 5.6
Furnace heat release ($10^3 Btu/ft^3$-hr)	10	10-13	15 – 20
Furnace gas residence time (sec)	3-4	2-4	2 – 3
Typical combustion temp (°C)	1090	1150	1260
Typical combustion temp (°F)	2000	2100	2300

Source: [15]

the alkali metals and alkali earth elements—can be a significant problem. The problem is exacerbated when chlorine is present, resulting in the formation of alkali chlorides. This problem is not severe at all when sawdust or hog fuel from primary wood processing is used, unless that wood comes from salt-water stored logs. It is more severe with woody materials such as vineyard or orchard prunings are used as fuel, since these materials come from biomass grown in an agricultural setting. It can also be more severe with urban wood waste, as suggested previously by the data in Table 4-9.

There is also consideration regarding whether woody biomass can poison or deactivate catalyst material for selective catalytic reduction (SCR) systems employed to reduce NO_x emissions. Mechanisms proposed include poisoning with arsenic, and deactivation due to high concentrations of calcium. While this consideration is normally raised in the context of cofiring systems, it is potentially more problematic with dedicated wood-fired boilers using silviculturally-induced fast growing species where the alkali and alkali earth concentrations in the flyash are relatively high. The low total ash concentrations in woody biomass tend to mitigate against such catalyst deactivation and poisoning. At this point in time the direct data concerning the SCR catalyst deactivation and poisoning is largely related to herbaceous biomass as will be discussed in Chapter 5. It is inferred to woody biomass.

4.3.2. Limitations on Dedicated Woody Biomass-fired Boilers

Commercially, dedicated woody biomass boilers are highly suited to the forest products industry, where the fuel is generated on-site and where both process steam and electricity—or shaft power—are required. However the fact that woody biomass does not pulverize readily, as does coal, limits the firing systems to stokers and CFB boilers. The plant site logistics and low bulk density of woody biomass typically limits the boiler to <88.3 kg/sec ($700x10^3$ lb/hr) of steam. This, in turn, has placed economic limits on the steam conditions associated with woody biomass boilers. At the maximum, steam conditions in woody biomass boilers are typically 100 bar/510°C (1450 psig/950°F); typical conditions for large-scale woody biomass boilers are on the order of 58.6 – 86.2 bar (850 – 1250 psig) and 440 – 510°C (825 – 950°F). These flows, pressures, and temperatures do not economically support the use of reheat cycles in dedicated woody biomass-fired boilers [15].

Because woody biomass boilers are not installed with reheat cycles, the electricity generating efficiencies of dedicated wood-fired systems are significantly lower than those associated with coal-fired central generating stations. The typical wood-fired generating station has a net station heat rate (NSHR), at best, of 13.7 MJ/kWh (13,000 Btu/kWh). Typically, the modern wood-fired generating stations employing 86.2 bar/510°C steam generate electricity with a NSHR of 14.8 MJ/kWh (14,000 Btu/kWh) or higher [15, 53]. This compares to the efficiencies of modern single and double reheat coal-fired boilers of 10.0 – 11.1 MJ/kWh (9,500 – 10,500 Btu/kWh). Modern natural gas-fired combined cycle combustion turbine generating stations can exhibit even more favorable NSHR values—on the order of 7.9 MJ/kWh (7,500 Btu/kWh) or better [54].

The scale of woody biomass boilers and associated generating equipment, as dictated by the maximum steaming capacity, is also considerably smaller than the scale of central station fossil fired boilers. Typically the largest wood-fired boilers used in electricity generation such as the McNeill Generating Station in Burlington, VT and the Kettle Falls Generating Station outside of Spokane, WA are on the order of 50 MW_e in capacity. The largest wood-fired boilers, if dedicated to the generation of electricity, would support some 70 – 75 MW_e of capacity. Typical woody biomass boilers installed throughout the USA have been

11 – 30 MW_e in capacity. This contrasts with modern coal-fired generating stations averaging 550 MW_e per boiler or more for all units installed in the USA since 1971 [54]. The consequence of this scale difference is reflected in operating and maintenance labor requirements. Woody biomass boilers typically generate 0.5 – 2 MW_e/employee, with the largest units such as McNeil Station generating 2.3 MW/plant employee and 1.14 MW/total employees on the payroll [52]. Coal-fired generating stations can generate an order of magnitude more electricity per employee. Modern natural gas-fired combined cycle combustion turbine-based stations also can be built at scales of 500 MW_e and greater. Economics do not favor stand-alone woody biomass electricity generating stations in modern industrial economies, except in niche applications or when incentives are applied.

4.4. Woody Biomass Cofiring in Pulverized Coal Cofiring Boilers

Recently, woody biomass has been tested and used as a supplementary opportunity fuel in large-scale utility boilers generating >126 kg/sec ($1x10^6$ lb/hr) high pressure/high temperature steam for use in electricity generating stations. Typical steam conditions employed in the generating stations where cofiring has been deployed generate steam on the order of 124 – 166 bar/538 – 551°C/538 – 551°C (1800 – 2400 psig/1000 – 1025°F/1000 – 1025°F). Some supercritical boilers have been employed in cofiring testing as well [20, 23].

Both pulverized coal (PC) and cyclone boilers have employed woody biomass cofiring. Among PC boilers, woody biomass—either sawdust or urban wood waste—has been cofired in the 140 MW_e Albright Generating Station boiler #3 of Allegheny Energy Supply Co., LLC., the Seward Generating Station of GPU Genco (now Reliant Energy), the Shawville Generating Station of GPU Genco (now Reliant Energy), the Greenidge Station of New York State Electric and Gas Co. (now AES), the Kingston and Colbert Fossil Plants of TVA, Plant Kraft and Plant Hammond of Southern Co. and several other units [20, 23, 55]. The experiments by Southern Co. at Plant Hammond and Plant Kraft were among the original tests of cofiring wood waste with coal [56].

In addition to PC and cyclone boilers, cofiring in circulating fluidized bed boilers will be a significant part of the renewable program of JEA (formerly Jacksonville Electric Authority) in Jacksonville, FL [57]. Co-gasification of woody biomass in an IGCC installation has been tested by Tampa Electric Company at its Polk Power Station Unit #1 [58], and gasification-based cofiring in a PC boiler has been commercialized in Lahte, Finland [59].

Cofiring in utility-owned PC and cyclone units capitalizes upon the large capacity, the more efficient reheat generating cycles, the more effective use of labor, and the generally significantly improved economics associated with large central station power plants. Cofiring of woody biomass in such coal-fired boilers, then, capitalizes upon the overall infrastructure and traditions of the generating stations currently in place. Cofiring woody biomass with coal also provides benefits to these plants in terms of improved environmental performance, with particular attention to reduced airborne emissions. In some cases, cofiring of woody biomass also provides specific operational benefits to the coal-fired generating stations as well.

Cofiring woody biomass in coal-fired boilers is performed differently depending upon the type of boiler; in PC boilers, woody biomass is typically injected separately from the coal unless very low percentages of woody materials are used (e.g., <5 percent by mass, or <2 percent by heat input). This separate injection is based upon adding flexibility to the boiler operations. Demonstrations have shown that separate injection permits concentrating the woody biomass in a single location to optimize NO_x reduction. Additionally, separate injection facilitates recovery of boiler steaming capacity that can be lost when wet coal is received by the power plant. In cyclone boilers the woody biomass is typically blended with the coal and introduced into the boiler with the coal [20]. In order to evaluate these uses of woody biomass as an opportunity fuel, case studies are employed as described below.

Because woody biomass must be reduced to <6.35 mm (<¼") particle sizes, several strategies have been developed. Some demonstrations have ground all wood to this particle size, while others have purchased sawdust and ground only the oversized particles. In all PC cofiring cases, considerable attention has to be paid to fuel preparation—and keeping the grinding of biomass separate from the grinding of coal.

4.4.1. *Cofiring Woody Biomass Fuel at the Albright Generating Station*

The Albright Generating Station cofiring project has been well described in the literature [20, 23, 48, 60]. The firing of sawdust with coal at the Albright Generating Station was initiated in 1999. Albright Generating Station boiler #3 is a 140 MW_e (net) tangentially-fired pulverized coal boiler equipped with a close-coupled overfire air (CCOFA) system and a separated overfire air (SOFA) system—both installed for NO_x reduction. The SOFA system has 3 rows of dampers, each of which can be opened from 0 – 100 percent. The unit burns a Pittsburgh seam medium volatile bituminous coal available locally. Typically, at full load, the unit fires about 54 tonne/hr (60 ton/hr) of coal.

The sawdust cofiring system was designed and installed in 2000 – 2001, and tested extensively immediately after that time. The system consists of a walking floor unloading conveyor, a disc screen to ensure that the particles fired in the unit are 6.35 mm x 0 mm (¼" x 0"), a 2-stage grinder to reduce oversized particles to the desired particle size, an agricultural silo for fuel storage, a reclaim system, a weigh belt feeder to meter the sawdust feed, and compressed air assisted rotary airlocks with blowers to feed and transport the sawdust to the boiler. The system is housed in a small metal building as shown in Figure 4.10.

The Albright #3 boiler is a single furnace unit with 4 elevations of burners or injectors in each corner. Sawdust injectors or burners were added to the unit in 2 corners between the B and C rows of coal burners or injectors. The sawdust injectors can follow the burner tilts associated with the coal burners. Sawdust is fed to the unit at up to 10 percent on a mass basis, or about 4.5 percent on a heat input basis. It is injected directly into the center of the fireball with a very high fuel/air ratio in the transport system. The stoichiometric ratio of the sawdust/transport air mix, when firing at 10 percent sawdust, is fuel-rich.

Results of over 500 hours of detailed testing have shown that cofiring woody biomass with coal at the Albright Generating Station created significant benefits for the plant. Operationally the cofiring of sawdust did not increase induced draft fan amps, or cause the plant to lose capacity in any way. In reality, the system permits the plant to compensate for difficulties and lost capacity associated with wet coal conditions that can be experienced in winter months.

Figure 4.10. The Albright Generating Station Cofiring Installation Receiving Sawdust from a Walking Floor Van

The efficiency losses associated with the sawdust cofiring were very small—typically <37 kJ/kWh (or <35 Btu/kWh). The efficiency losses were associated first with the hydrogen content of the fuel and the moisture content of the fuel. There was a 4.5°C (8°F) increase in the temperature of the flue gas exiting the air heater as a consequence of less air being required for coal combustion; some combustion air entered the boiler as transport air for the sawdust. These effects, however, were minor [48, 60].

The environmental benefits associated with biomass cofiring at the Albright Generating Station have been substantial. Firing 10 percent sawdust—or 5.4 tonne/hr (6 ton/hr)—results in a reduction of SO_2, NO_x, and mercury simultaneously. At the same time it does not result in an increase in opacity, particulate emissions, or CO emissions. Further the reduction in NOx does not come with an increase in unburned carbon in the flyash [48].

The environmental benefits of SO_2 and mercury emissions reductions are the direct result of the fuel characteristics as discussed previously in Section 4.2.2 and 4.2.4. The woody biomass is virtually free of both sulfur and mercury as previously shown. The NO_x reductions are more dramatic, and more complex.

NO_x emissions reductions occurred as a function of cofiring, and the integration of the biomass injection system with the SOFA system. The following regression equation has been developed to describe the NO_x reduction phenomenon ($r^2 = 0.87$) [47, 58]:

$$NO_x = 0.097+3.19\text{x}10^{-5}(H)-0.004(W_t)+0.013(EO_2)-0.0002(SOFA) \quad [4\text{-}4]$$

Where NO_x is measured in kg/GJ of heat input, H is heat input in GJ/hr, W_t is the sawdust feed rate in tonnes/hr, EO_2 is the excess O_2 measured in the control room (total basis), and SOFA is the total of the three levels of SOFA damper positions expressed as a percentage. This equation, with 67 degrees of freedom, has a probability of random occurrence, rather than as a consequence of the variables, of $2.49\text{x}10^{-27}$. The intercept has a probability of random occurrence of 0.045. The heat input has a probability that its influence is random of 0.209; it may not be a significant variable in this boiler. The probability that the influence of sawdust input is a random occurrence is $6.6\text{x}10^{-6}$. The probabilities that the influence of excess O2 and SOFA damper positions are random occurrences are 0.00098 and $5.26\text{x}10^{-20}$ respectively.

The English equivalent of this equation is as follows:

$$NO_x = 0.225+7.83\text{x}10^{-5}(H)-0.001(W_t)+0.03(EO_2)-0.005(SOFA) \quad [4\text{-}5]$$

Where NO_x is measured in lb/10^6 Btu, H is measured in 10^6 Btu/hr, and W_t is measured in tons/hr.

The mechanism employed at the Albright cofiring program involves flooding the center of the fireball with volatiles, creating a fuel-staging condition at that point. This enhances the effect of the SOFA system. Further, because of the volatility of the biomass, the fuel staging does not come at the expense of unburned carbon in the flyash—or loss on ignition (LOI). The NO_x emissions were reduced to 108 g/GJ (0.25 lb/10^6 Btu) without significantly reducing boiler efficiency, and without increasing LOI [48, 60].

During a recent outage, the boiler was inspected to determine whether the practice of cofiring caused any increase in slagging or fouling deposits. The evidence was clear that, after significant cofiring operating hours at the Albright Generating Station, no abnormal buildup of deposits occurred. This was consistent with evaluations of sootblowing schedules during the test periods [61].

Cofiring was sufficiently successful at the Albright Generating Station that Allegheny Energy Supply Co., LLC—the owner of this generating station—chose to continue the practice beyond the initial demonstration program. The 100 hr test of the facility demonstrated its reliability while the environmental benefits proved sufficiently economically attractive to continue the process.

4.4.2. Cofiring Sawdust at the Seward Generating Station

The benefits obtained at the Albright Generating Station in a tangentially-fired PC boiler essentially mirrored those benefits previously demonstrated at the Seward Generating Station in a front-fired boiler. This demonstration has been previously described in numerous papers and reports [20, 23, 62 - 64].

The Seward Station, owned at the time of demonstration by GPU Genco (now owned by Reliant Energy) operated three boilers: two 32 MW_e wall-fired or front-fired boilers and a 147 MW_e tangentially-fired boiler. Before boiler #12, one of the 32 MW_e boilers was removed from continuous service in favor of a repowering program, GPU Genco undertook a sawdust cofiring demonstration.

Boiler #12, which was selected for the cofiring testing, is a 1950 vintage Babcock and Wilcox (B&W) front wall-fired boiler, with a capacity of approximately 37.7 kg/sec (300,000 lb/hr) of 45.9 bar/446°C (675 psig/835°F) steam. Along with Boiler #14, a twin to #12 except it has been modified with low NOx burners, it feeds steam to a common header that, in turn, feeds a 64 MW (net) Westinghouse steam turbine (Unit #4). The net heat rate for these units is approximately 14.98 MJ/kWh (14,200 Btu/kWh). The third boiler (#15) is a 147 MWe (gross) Combustion Engineering (CE) boiler built in 1957 which is a tangentially-fired pulverized coal boiler.

The demonstration was initiated by a series of parametric tests employing separate injection of the sawdust into the coal-fired boiler, and maintaining the integrity of the pulverized coal preparation and delivery system. During these parametric tests, Boiler #12 was equipped with a biomass surge bin, metering augers, lock hoppers, and transport pipes. The transport pipes connected the biomass delivery system to the unused centerpipes of the three top burners. Each burner was equipped with a separate metering auger, a separate lock hopper, and a separate blower. Each such system could deliver 2.72 tonne/hr (6,000 lb/hr) of sawdust to the boiler. As a practical matter, however, the unit was typically operated in the 2-tonne/hr range. The demonstration project consisted of co-firing sawdust with pulverized coal in a 32 MWe wall-fired pulverized coal boiler (#12) by utilizing separate injection of the wood at a rate of approximately 2 tonnes per hour (up to 10 percent on a heat basis) for an extended period of time.

The initial tests were conducted with green sawdust (39 percent moisture), old sawdust (49 percent moisture), and dry sawdust at 14 percent moisture. Cofiring levels ranged from 0 percent to 18 percent on a mass basis or up to about 10 percent on a heat input basis. These tests demonstrated that there would be no deleterious capacity impacts from cofiring, that efficiency losses could be modest and manageable, and that emissions impacts would be beneficial.

When the parametric tests were combined, they yielded an efficiency regression equation as follows ($r^2 = 0.89$)

$$\eta = 87.6 - 0.14(EO_2) - 0.16(UBC) - 0.11(W) \qquad [4\text{-}6]$$

Where EO_2 is percent excess O_2 measured on a wet basis, UBC is percent unburned carbon in the flyash, and W is percent wood in the fuel mix on a heat input basis. The probabilities of the equation, or any component, being random are as follows: total equation, 0.000125; intercept, 4.24×10^{-24}; EO_2, 0.77; UBC, 0.0027; and W, 0.00072. Cofiring reduced boiler efficiency by about 1 percent for every 10 percent on a heat input basis—or 1 percent for every 20 percent on a mass input basis [61-62].

The tests also demonstrated that carbon conversion—combustion efficiency—was not degraded by cofiring. CO measurements showed a range of 8.5 ppmv to 18 ppmv when cofiring percentages ranged from 0 to 20 (mass basis). Benefits demonstrated during the tests included the ability to recover some capacity lost due to the impacts of wet coal on the pulverizer. These benefits were of particular economic significance [16].

NO_x emissions reductions were pronounced at the Seward Generating Station parametric tests, as is shown in Figure 4.11. Again the driving variable, determined statistically, was the fuel volatility, as shown in the Albright tests as well. The success of the parametric tests led to the construction of a demonstration facility. This demonstration facility consisted of a walking floor unloader, capable of receiving sawdust from a walking floor truck. This unloader fed a trommel screen producing a biomass fuel with a top particle size of 6.25 mm. The screened sawdust was then stored in a silo. When reclaimed from the silo it was transported across a weigh belt feeder to the feed screws and rotary airlocks. It was then injected into the centerpipe of the middle burner in each row of burners.

The results of the demonstration were as favorable as those obtained during the parametric tests. Capacity was not reduced as a function of cofiring. Efficiency reductions were modest. Operationally there were no problems with opacity, unburned carbon or LOI in the flyash, or CO emissions. The Seward Station demonstration also demonstrated an operational benefit—the ability of cofiring to help

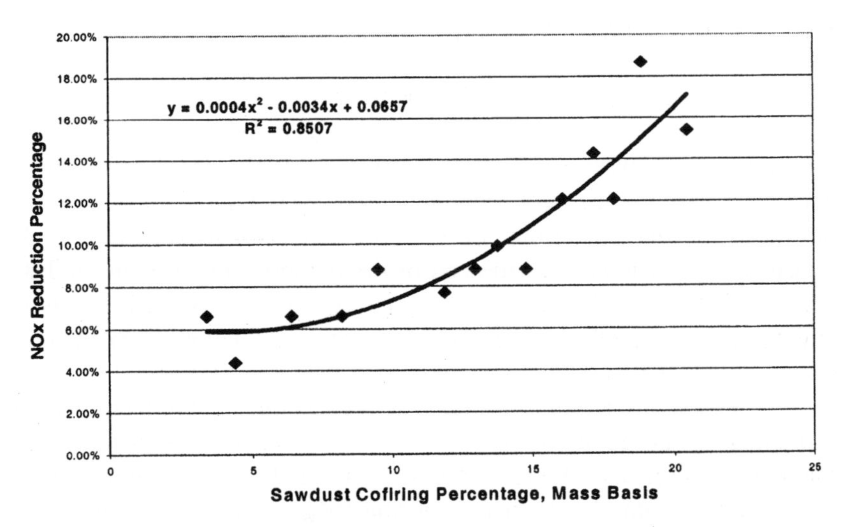

Figure 4.11. NO_x Reduction During the Parametric Testing at the Seward Generating Station

Source: [23]

recover capacity lost during wintertime conditions and the firing of wet coal. In one parametric test it was demonstrated that 10 percent cofiring (heat input basis) could be used to recover 12.5 percent of the steaming capacity of the boiler when this capacity was lost to wet coal and its impact upon pulverizer performance.

The most dramatic results again were in the area of NO_x reduction as shown in equation [4-7] and Figure 4-12. The data on NO_x emissions resulted in the following regression equation ($r^2 = 0.93$):

$$NO_x = 18.92 - 647.4\,(W_m) + 9.66(L) + 59.9(EO_2) \qquad [4\text{-}7]$$

Where NO_x = oxides of nitrogen, ppmvd at 3% O_2 (dry basis), L = load measured as main steam flow in kg/sec, EO_2 = excess O_2 reported in the control room (total basis), and W_m = wood cofiring percentage, mass basis. The equation is quite robust. The probability that the results are a random occurrence is $4.3\text{x}10^{-6}$, the probability that the W_m term is a random event is $8.3\text{x}10^{-7}$, the probability that the L term occurs randomly is $2.1\text{x}10^{-5}$, and the probability that the EO_2 term occurs randomly is $2.3\text{x}10^{-5}$. The impact of cofiring sawdust at the Seward Generating Station Boiler #12 wall-fired boiler demonstration is also shown in Figure 4-12. Note the linearity of NO_x reduction as a function of cofiring percentage [61 - 64].

The key to the NO_x reduction at the Seward Generating Station, like that of the Albright Generating Station, came from flooding the combustion zone with highly volatile sawdust. In the case of the Seward demonstration, the burners were modified such that the sawdust was blown down the centerpipe of conventional burners with Eagle Air registers. The spreaders at the end of the burners had an angle of 120°, creating a short, bushy coal flame. The sawdust diffuser angle was set slightly greater than the spreader angle, creating a volatile "flame within a flame" at the base of the coal flame.

Visual inspection showed that the sawdust ignited rapidly and promoted ignition of the coal in a fuel-rich environment. This phenomenon has caused consideration of design details to optimize multi-fuel burner designs for biomass cofiring [65]. Again the creation of a fuel-staging condition with volatile flooding caused the reduction in NO_x

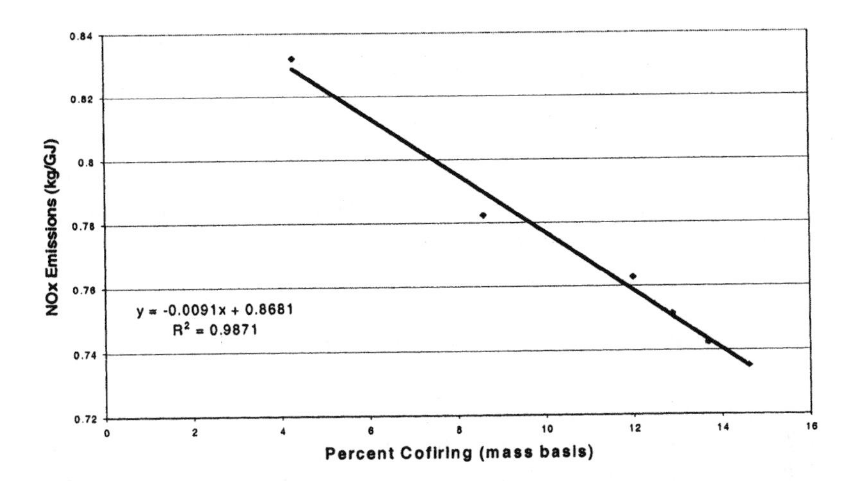

Figure 4-12. NO_x Emissions Measured During the Seward Generating Station Demonstration
Source: [23]

emissions without increasing LOI. The Seward Generating Station sale to Sithe Energies, and then Reliant Energy—coupled with the station repowering program—caused the termination of this demonstration.

4.4.3. Conclusions Regarding Cofiring Woody Biomass in PC Boilers

Woody biomass cofiring in PC boilers has the benefit of capitalizing upon the highly efficient designs of such systems with more severe steam conditions and with reheat cycles. By using cofiring, there is no limit to the capacity of the boiler per se—only a limit on the contribution of biomass to the generation from that boiler. Because woody biomass cofiring in PC boilers has proven successful, it has been installed and/or tested at such locations as the Greenidge Generating Station (AES), Colbert Fossil Plant (TVA), Kingston Fossil Plant (TVA), Plant Hammond (Georgia Power), Plant Kraft (Savannah Electric), and other installations as well [20, 23].

Benefits from cofiring woody biomass in PC boilers can readily be seen. While there is a modest decrease in boiler efficiency, there is no loss of boiler capacity. As was shown at Seward Generating Station, capacity lost to wet coal could be recovered by separate injection cofiring. The overwhelming benefits, however, were in the area of emissions reduction. SO_2, NO_x, and mercury emissions have been consistently reduced by cofiring woody biomass in coal-fired PC boilers. These reductions are dramatic and readily quantified as a function of capitalizing upon the fuel characteristics of woody biomass.

4.5. Cofiring Woody Biomass in Cyclone Boilers

Cofiring woody biomass in cyclone boilers has involved both sawdust and urban wood waste. Initial programs involved the King Station of Northern States Power outside of St. Paul, MN, cofiring sawdust from the Andersen Windows plant in a 600 MW_e supercritical boiler fired by 12 cyclone barrels. Cofiring wood waste in cyclone boilers also included the Allen Fossil Plant of Tennessee Valley Authority, the Bailly and Michigan City Generating Stations of Northern Indiana Public Service Co. (see Chapter 2 for a discussion of using urban wood waste with petroleum coke at Bailly Generating Station), the Gannon Station of Tampa Electric Company, and the Willow Island Generating Station of Allegheny Energy Supply Co., LLC. Two demonstrations are discussed below: The Allen Fossil Plant and the Willow Island Generating Station.

Unlike cofiring woody biomass in PC boilers, cofiring sawdust or urban wood waste in cyclone boilers virtually always involves blending the biomass with the coal on the way to the bunkers rather than separately injecting the sawdust into the boiler. Only the King Station system involves separate injection of dry woody biomass into the boiler. There have been studies of reburn using woody biomass in cyclone boilers [28 - 29], however this has not been commercially applied to date.

4.5.1. Cofiring at the Allen Fossil Plant

Cofiring at the Allen Fossil Plant of TVA has been described in a series of papers and reports [23, 33, 66-68]. The cofiring tests were conducted from 1993 – 1996, and included evaluating the impact of cofiring percentage, particle size of the wood fuel, excess O_2, and air distribution within the cyclone barrels as they impacted system performance.

The Allen Fossil Plant, located in Memphis, TN, houses 3 balanced draft cyclone boilers, each having a capacity of 272 MW_e (net). Each boiler is equipped with 7 cyclone barrels—3 on the north side and 4 on the south side. The cyclone barrels are on a single elevation. This plant is located in a wood products oriented area, with numerous sawmills, furniture mills, and flooring mills in the immediate vicinity. In tests cofunded by TVA and EPRI, cofiring woody biomass in cyclone boilers was extensively tested and evaluated at this location.

Cofiring percentages up to 20 percent by mass (10 percent by heat input) were tested at the Allen Fossil Plant. Sawdust was obtained from local sources. Initially both green and dry sawdust was obtained; the best success was obtained with green sawdust at about 40 percent moisture, and dry sawdust was abandoned because of handling problems. Sawdust was received in the coal yard and screened using a large trommel screen. The sawdust was then stocked out in the coal yard and reclaimed for blending with the base coal on the belts leading to the bunkers. Base coals employed included both Illinois basin coal and Utah bituminous coal. Experiments were conducted varying the particle size of the biomass from 6.25 mm (¼" x 0") to 38 mm (1½" x 0"). Experiments were conducted varying the excess O_2 percentage from 2.2 percent to 3.5 percent. Some experimentation was conducted varying the primary air/secondary air ratio, and other experiments included mixing tire-derived fuel chips with the sawdust and coal [33, 66-68].

The Allen Fossil Plant tests set the stage for all subsequent tests in cofiring, and documented the ability of this technology to reduce fossil CO_2 emissions along with SO_2 emissions. Further, these tests documented the ability of cofiring to achieve such reductions with modest efficiency penalties.

The initial tests documented that cofiring could occur without causing a loss in boiler capacity, a problem predicted initially. There was in increase in feeder speeds, but there was sufficient feeder capacity to accommodate the cofiring. There was a loss in boiler efficiency, however there was no need to increase excess O_2. Further, there was no increase in air heater in-leakage or air heater exit temperature as a function of biomass cofiring [66].

The testing at the Allen Fossil Plant also demonstrated that an interesting phenomenon occurs when cofiring woody biomass with coal in cyclone boilers: the combustion temperature does not decrease appreciable, however the furnace exit gas temperature does decrease measurably. Modeling by Reaction Engineering International confirmed that the cyclone barrel experiences no decrease in cyclone barrel temperature. Figure 4-13 shows the results of modeling a blend of 15 percent sawdust (mass basis) with Utah bituminous coal in the cyclone barrels at Allen Fossil Plant. The temperature profile virtually mirrors that associated with the modeling of coal alone. This further indicates that the operational impacts of cofiring woody biomass in cyclone boilers is inconsequential [33].

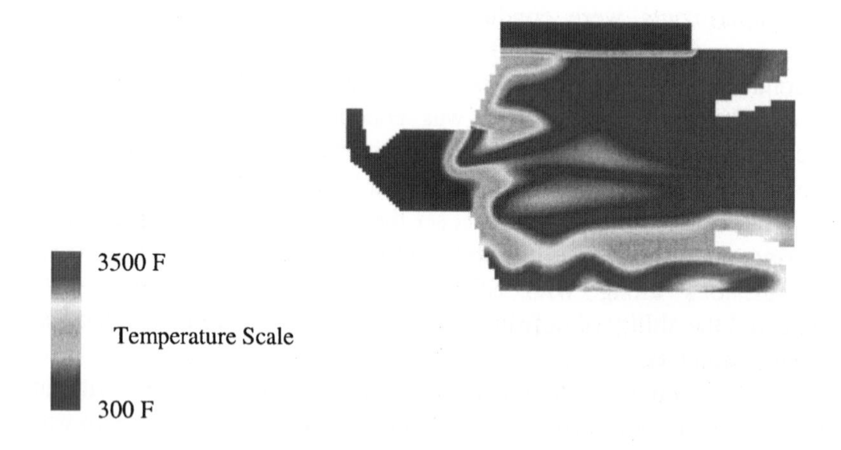

Figure 4-13. Initial Temperature Modeling of an Allen Fossil Plant Cyclone Cofiring 15 Percent Sawdust with 2.6 Percent Excess O_2

Source: [33]

The temperature phenomenon described above was consistent with measurements of slagging and fouling. During many of the tests sootblowing was not permitted, and the evaluation focused upon the rate of temperature increase at the economizer exit as a function of time. In virtually all cases the rate of temperature rise was faster when no biomass was in the fuel mix than when biomass was incorporated in the fuel mix. The rate of temperature rise measures deposition indirectly, since it measures reduction in heat transferred from the combustion products to the steam as a function of fouling.

The tests were most significant, however, in documenting the ability of sawdust cofiring to reduce NO_x emissions. When cofiring with <6.25 mm particles, tests at the Allen Fossil Plant resulted in the ability to reduce NOx emissions conforming to the following equation:

$$NO_x = 0.000423(FR)+0.0904(EO_2)+1.554(FN)-0.629(V/FC)-0.752 \quad [4-8]$$

Where NO_x is measured in kg/GJ, FR is firing rate in GJ/hr, EO_2 is excess O_2 expressed as a percentage (total basis), FN is fuel nitrogen expressed in kg/GJ, and V/FC is the volatile/fixed carbon ratio from the proximate analysis of the blend of coal and woody biomass. Alternatively:

$$NO_x = 0.001(FR)+0.210(EO_2)+1.554(FN)-1.46(V/FC)-1.748 \quad [4-9]$$

Where NO_x is expressed in lb/10^6 Btu, FR is firing rate expressed in 10^6 Btu/hr, and FN is fuel nitrogen expressed in lb/10^6 Btu. The biomass tested had a low fuel nitrogen content (typically 0.1 – 0.3 percent) and a very high V/FC ratio (typically >4.0) particularly when compared to coal. Consequently it drives the results downward. The coefficient of determination (r^2) for this equation—in SI or English measure—is 0.86. The probabilities that these terms are random occurrences are as follows: Equation, 5.14×10^{-8}; FR, 0.108; EO_2, 0.090; FN, 0.006; V/FC, 0.0002. These data suggest a very robust equation. The volatility influence is particularly important, as shown in Figure 4-14.

Tests at the Allen Fossil Plant also confirmed that particle size had a significant influence on the ability to reduce NO_x emissions. As particles became larger, the NO_x reduction decreased. When particles reached 25 mm in size, there was no reduction in oxides of nitrogen.

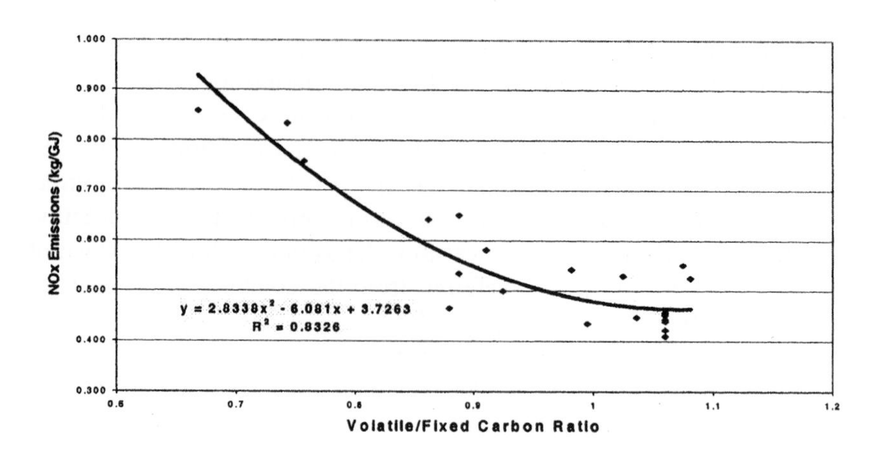

Figure 4-14. The Influence of Fuel Volatility on NO_x Emissions at the Allen Fossil Plant Cofiring Tests
Source [66]

Under such conditions diffusion effects reduced the practical volatility of the biomass in the cyclone barrel [33].

Sulfur emissions also were reduced as a function of substituting a sulfur-free fuel—sawdust—for coal. Sulfur emissions reductions were more dramatic when cofiring against Illinois basin coal than Utah coal due to the differences in the sulfur content of the fossil energy source. The reduction in emissions documented the value of this cofiring system.

The cofiring tests at Allen Fossil Plant led directly to cofiring testing at the Michigan City Generating Station boiler #12 of NIPSCO [69-70], the Bailly Generating Station of NIPSCO as described previously in Chapter 2 (see, also, [71-72]), and the Willow Island Generating Station as discussed below.

4.5.2. Woody Biomass Cofiring at the Willow Island Generating Station

The biomass cofiring program at Willow Island Generating Station of Allegheny Energy Supply Co., LLC has demonstrated the commercial aspects of this practice. This program, funded jointly by Allegheny and the National Energy Technology Laboratory of US Department of Energy has been described extensively in conference papers and reports [20, 61, 73-75].

Willow Island Generating Station is located at the Pleasants-Willow Island plant site of Allegheny on the Ohio River, just north of Parkersburg, WV, is a major site for power generation. Willow Island Boiler #2 is a 188 MW_e (net) pressurized cyclone boiler equipped with 5 cyclone barrels installed on the front wall of the boiler. The unit is equipped with a "hot side" electrostatic precipitator for particulate control. It is equipped with a separated overfire air (SOFA) system, although that system is not commonly used.

The Pleasants-Willow Island location is in the heart of a wood products region, and numerous sawmills and secondary products manufacturing firms are located within economic haul distance to this plant. Consequently it is ideally suited for cofiring woody biomass. In the year 2000, Foster Wheeler designed a system for blending up to a maximum of 15 percent (mass basis) sawdust with coal at this plant. From an operational perspective, the system is designed for 10 percent cofiring on a mass basis as a long-term target regime [73]. Previously the plant had installed a system for cofiring TDF with coal at this location.

The system designed for Willow Island begins with a receiving station where walking floor vans discharge sawdust into a below-grade hopper. The sawdust is then conveyed to a 50 tonne/hr disc screen producing <9.53 mm (3/8") particles for cofiring. Oversized particles are directed to a grinder for reduction to particle sizes consistent with cyclone firing. The biomass fuel particles so prepared are then conveyed to a large (400 tonne capacity) walking floor bin where they are stored under roof until reclaimed for use. The sawdust is reclaimed by counter-rotating twin augers and fed to a metering belt. The metering belt then moves the sawdust to the main belt conveying washed Pittsburgh seam coal to the bunkers. The TDF is introduced separately into the system [75]. The blended fuel is processed through secondary crushers on the way to the cyclone barrels. The blended fuel is then fed by Stock feeders to the cyclone barrels. Figures 4-15 and 4-16 depict this installation.

The project was constructed in 2001 and early 2002. Testing has commenced. As of this writing some 5,000 tonnes of sawdust has been burned at the Willow Island Generating Station. Cofiring was

Figure 4-15. The Installation at Willow Island Generating Station Receiving Sawdust

Figure 4-16. The 400 tonne Sawdust Storage Bin Under Construction at Willow Island Generating Station

demonstrated at ratios of up to 10 percent sawdust (mass basis) and up to 5 percent TDF (mass basis) [76].

Operationally the testing has proven successful. Cofiring has occurred at loads from 106 MW_e (net) to 188 MW_e (net) without compromising boiler capacity. Cyclone feeder speeds have increased as a consequence of cofiring a modest calorific value biomass fuel with a typical heat content of 11.14 – 11.60 MJ/kg (4800 – 5000 Btu/lb) with a high calorific value coal with a typical heat content of about 29.0 MJ/kg (12,500 Btu/lb).

Efficiency losses have been extremely modest, with no observed increase in the required excess O_2, no increase in air heater exit temperature, and no change in air heater in-leakage [61].

The temperature phenomena experienced at Allen Fossil Plant have again been experienced at the Willow Island Generating Station. There has been no decrease in cyclone boiler temperature as observed, and as calculated by extensive modeling. At the same time there has been a noticed decrease in furnace exit gas temperature (FEGT). This decrease in FEGT, of >110°C (200°F) has created benefits for the plant. At the same time this FEGT decrease has not caused any loss of main steam or reheat steam temperature. Cofiring has not caused any increase in LOI [61]. This phenomenon was also observed and measured at Michigan City Generating Station when cofiring urban wood waste with a blend of western bituminous and PRB coals. It was also modeled for the cofiring tests at Bailly Generating Station.

Emissions reductions have occurred at the Willow Island Generating Station as a consequence of biomass cofiring. SO_2 and mercury emissions reductions have occurred as a consequence of substituting a sulfur-free and mercury-free fuel for coal.

Cofiring has not caused opacity excursions or increases in particulate emissions. There have been some modest reductions in NO_x emissions, although they have not been as pronounced as in the cofiring at PC locations or as in the cofiring at the Allen Fossil Plant. Modeling suggests that about a 10 percent reduction in NO_x emissions can be expected over the long term when cofiring at 10 percent on a mass basis or 4.5 percent on a heat input basis, and the testing and operational activities are indicating progress towards this expected result of cofiring.

4.5.3. *Conclusions Regarding Cyclone Boiler Cofiring of Woody Biomass*

Woody biomass can readily be cofired in cyclone boilers as has been demonstrated in numerous locations. Because cyclone boilers do not utilize a pulverized fuel, the biomass can be blended with the coal going to bunkers and then the blend can be fired directly in the cyclone barrels.

Like the PC boilers, the cyclone boilers experience no capacity impacts associated with cofiring. Although they experience increased feeder speeds, such feeder limitations are rare in cyclone firing. Efficiency impacts are minor. Environmental benefits are significant.

Like cofiring in PC boilers, the evidence is clear that cofiring does not increase deposition of inorganic matter in the boiler—either as slagging or as fouling. Like cofiring in PC boilers, and firing woody biomass in dedicated combustion systems, there is no direct evidence of the effect of woody biomass firing on SCR catalyst poisoning or deactivation. However there remain questions resulting from tests with herbaceous biomass materials. With cofiring in coal-fired boilers, the presence of 10 – 20 percent biomass on a mass basis has the net effect of reducing the ash loading in kg ash/GJ of fuel (or lb ash/10^6 Btu of fuel). While it is projected that the influence of cofired woody biomass on SCR catalyst would be minor if at all, that remains to be proven by research which is now on-going [76].

4.6. Conclusions Regarding Using Woody Biomass as an Opportunity Fuel

Woody biomass, probably the first combustible fuel used by mankind, remains an opportunity fuel of choice for many installations. It is the dominant fuel of the forest products industry and, in that, contributes some 3.5 EJ (or 3.5×10^{15} Btu) per year to the US economy. It contributes proportionately more to the Scandinavian economies as well.

Typically wood fuels—including spent pulping liquor—have been used in dedicated boilers. These boilers in the pulp and paper industry commonly support cogeneration applications, generating 58.6 – 100 bar

(850 – 1450 psig) and 450 – 540°C (850 – 1000°F) main steam and exhausting low pressure 3.5 – 10.3 bar (50 – 150 psig) steam from backpressure or automatic extraction turbines for use in process applications. Woody biomass supports the generation of over 7,000 MW of electricity for the US economy, and considerable electricity for northern European economies as well. In northern European economies, combined heat and power (CHP) applications are commonly employed, producing both electricity and district heat for communities.

Dedicated wood fired boilers are typically capacity limited by plant logistics as well as by volumetric heat release rates. The largest of these boilers typically generate 75.7 – 88.3 kg/sec (600,000 – 700,000 lb/hr) of steam for turbine and process applications. These capacity limitations have economically precluded the construction of reheat boilers being fired by woody biomass. Consequently their efficiency in the generation of electricity through condensing turbine applications is limited; typical net station heat rates (NSHR) for the largest and most efficient wood fired power plants are on the order of 13.7 – 14.8 MJ/kWh (13,000 – 14,000 Btu/kWh). NSHR values for many wood-fired power plants are on the order of 18.4 MJ/kWh (17,500 Btu/kWh).

More recently, cofiring woody biomass in PC and cyclone boilers used by electricity generating utilities has been employed selectively as a means for capitalizing upon the improved efficiencies of large, high pressure, reheat boilers. Utility boilers employing cofiring—either in test or in commercial deployment applications—have ranged in size up to 600 MW_e, and have commonly been in the 140 – 270 MW_e range.

In deploying woody biomass in cofiring applications, utilities have found that the fuel characteristics of this class of fuels can be used advantageously. The high reactivity is most beneficial. The low nitrogen content of sawdust and the low sulfur and mercury contents of all woody biomass fuels are also most beneficial.

The use of woody biomass in cofiring applications has not been shown to cause a decrease in boiler capacity. In PC firing, it has been shown that cofiring can help recover lost capacity when the unit is pulverizer limited and firing wet coal under winter conditions. Woody biomass cofiring does cause modest efficiency losses, however these are readily managed. Operationally, most tests have shown that base combustion temperatures have not been compromised, however FEGT values have decreased. At the same time main steam and reheat steam temperatures have not decreased.

Cofiring woody biomass has reduced airborne emissions resulting from coal combustion in central station power plants. It has been shown that every tonne of woody biomass fired reduces the emission of fossil CO_2 by >1 tonne directly, and by about 3 tonnes of total fossil CO_2 equivalent [20, 23]. Because of the low sulfur and mercury content of woody biomass, SO_2 and mercury emissions are also decreased. NO_x emissions are reduced as a function of the fuel volatility, the low nitrogen content when firing sawdust, and by reduced FEGT values. Further, the reduction in oxides of nitrogen occurs without LOI penalty. Cofiring can be integrated with low NO_x SOFA systems as well, further decreasing these emissions. Figure 4-17 presents a compilation of test data from USDOE-sponsored demonstrations.

If the influence of cofiring were strictly the substitution of a low nitrogen fuel for coal, then the data points should be on the line indicated.

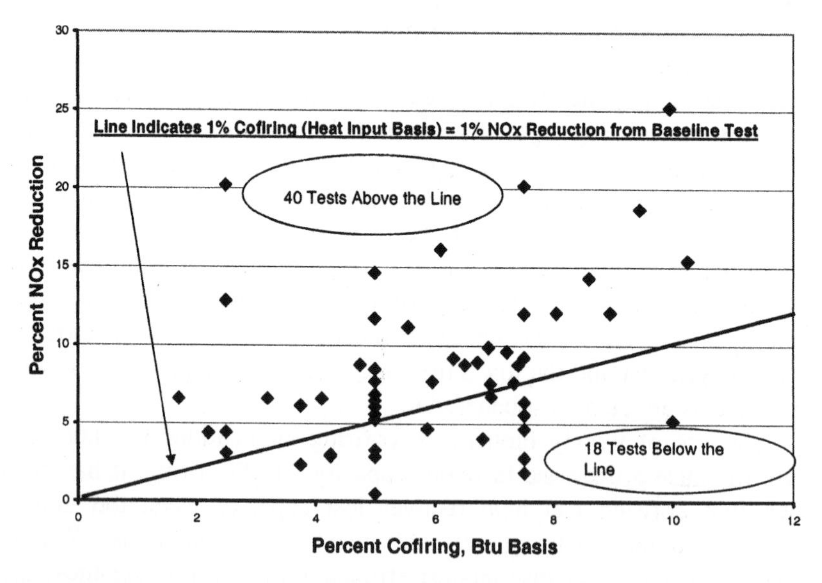

Figure 4-17. NOx Reduction as a Function of Cofiring—Summary of all USDOE-Sponsored Demonstrations.
Source [20]

The reality is that there is significant evidence that cofiring woody biomass with coal can dramatically reduce NO_x emissions, particularly if the boiler operations are designed to achieve this objective.

In conclusion, then, cofiring of woody biomass with coal has shown itself as a significant new approach to the use of this opportunity fuel. Such an approach complements the use of woody biomass in dedicated boilers and related facilities. It provides additional markets for these energy sources in the coming years and decades.

4.7. References

1. Enzer, H., W. Dupree, and S. Miller. 1975. *Energy Perspectives: A Presentation of Major Energy and Energy Related Data.* US Department of the Interior. Washington, DC.
2. Tillman, D. 1978. Wood as an Energy Resource. Academic Press, New York.
3. Walker, J.E. 1966. Hopewell Village: The Dynamics of a Nineteenth Century Iron-Making Community. University of Pennsylvania Press, Philadelphia, PA.
4. National Materials Advisory Board. 1975. *Problems and Legislative Opportunities in the Basic Materials Industries.* National Research Council, National Academy of Sciences, Washington, DC.
5. Energy Information Agency. 2002. Annual Energy Review. US Department of Energy, Washington, DC.
6. Henry, J. 1979. The Silvicultural Energy Farm in Perspective. in Progress in Biomass Conversion, Vol 1. Academic Press, New York. pp. 215-256.
7. College of Forest Resources, University of Washington. 1979. Energy From Wood: A Report to the Office of Technology Assessment, Congress of the United States. Energy from Biological Processes Vol III—Appendixes. Office of Technology Assessment, Washington, DC.
8. Committee on Renewable Resources for Industrial Materials. 1976. The Potential of Lignocellulosic Materials for the Production of Chemicals, Fuels, and Energy. National Research Council, National Academy of Sciences. Washington, DC.

9. Tillman, D. 1982. The Cost of Electricity from Silvicultural Fuel Farm-Based Power Plants. in Energy from Forest Biomass (Smith, W., ed.). Academic Press, New York. pp. 253-274.
10. Ostlie, L.D. 1993. Whole Tree Energy Technology and Pilot Test Program. Proc. Strategic Benefits of Biomass and Waste Fuels. Conference. EPRI, Palo Alto, CA. March 30 – April 1. Washington, DC.
11. Robertson, T., and H. Shapouri. 1993. Biomass: An Overview in the United States of America. Proc. First Biomass Conference of the Americas. Burlington, VT. Aug 30 – Sep 2. pp. 1 – 17.
12. Hughes, E. 2000. Biomass cofiring: economics, policy and opportunities. BIOMASS AND BIOENERGY. 19(6):457-466.
13. Haygreen, J. and J. Bowyer. 1982. Forest Products and Wood Science. Iowa State University Press, Ames, IA.
14. Forest Products Laboratory. 1974. Wood Handbook: Wood as an Engineering Material. US Government Printing Office, Washington, DC.
15. Tillman, D. 1991. The Combustion of Solid Fuels and Wastes. Academic Press, San Diego, CA.
16. Prinzing, D. 1996. *EPRI Alternate Fuels Database.* EPRI, Palo Alto, CA. Report TR-107602.
17. Tillman, D.A. 1999. *Biomass Cofiring: Field Test Results.* Electric Power Research Institute, Palo Alto, CA. Report TR-113903.
18. Sjostrom, E. 1981. Wood Chemistry: Fundamentals and Applications. Academic Press, New York.
19. Wenzl, H. 1970. The Chemical Technology of Wood. Academic Press, New York.
20. Tillman, D.A. 2002. Cofiring Technology Review, Final Report. National Energy Technology Laboratory, US Department of Energy, Pittsburgh, PA.
21. Miller, B.G., S.F. Miller, C. Jawdy, R. Cooper, D. Donovan, and J. Battista. 2000. Feasibility Analysis for Installing a Circulating Fluidized Bed Boiler for Cofiring Multiple Biofuels and Other Wastes at Penn State University: Second Quarterly Technical Progress Report. Work Performed Under Grant No. DE-FG26-00NT40809.
22. Johnson, D.K. et. al. 2001. Characterizing Biomass Fuels for Cofiring Applications. Proc. Joint International Combustion Symposium. American Flame Research Committee. Kaui, Hawaii. Sep 9 – 12.

23. Tillman, D.A. 2001. Final Report: EPRI-USDOE Cooperative Agreement: Cofiring Biomass With Coal. Contract No. DE-FC22-96PC96252. EPRI, Palo Alto, CA.
24. Tillman, D., B. Miller, and D. Johnson. 2002. Nitrogen Evolution from Biomass Fuels and Selected Coals. Proc. The Pittsburgh Coal Conference. Sep 23-26. Pittsburgh, PA.
25. Johnson, D., D. Tillman, and B. Miller. 2003. Reactivity of Selected Opportunity Fuels: Measurements and Implications. Proc. Electric Power Conference. March 3-6, Houston, TX.
26. Baxter, L.L. et. al. 1996. Nitrogen Release during Coal Combustion. ENERGY & FUELS. 10(1): 188-196.
27. Brown, A., D. Dayton, and J. Daily. 2001. A Study of Cellulose Pyrolysis Chemistry and Global Kinetics at High Heating Rates. ENERGY AND FUELS. 15(5): 1286-1294.
28. Shafizadeh, F. and W. DeGroot. 1977. Thermal Analysis of Forest Fuels. in Fuels and Energy From Renewable Resources (Tillman, D., K. Sarkanen, and L. Anderson, eds). Academic Press, New York. pp. 93-114.
29. Harding, N.S. and B.R. Adams. 2000. Biomass as a reburning fuel: a specialized cofiring application. BIOMASS AND BIOENERGY. 19(6): 429-446.
30. Adams, B.R. and N.S. Harding. 1996. Modeling of Wood Reburning for NOx Control. Proc. 1996 International Joint Power Generation Conference. Oct 13-17. Houston, TX. ASME, New York. pp. 455-462.
31. Junge, D.C. 1975. *Boilers Fired with Wood and Bark Residues.* Research Bulletin 17. Forest Research Laboratory, Oregon State University, Corvallis, OR.
32. Junge, D.C. 1979. *Design Guideline Handbook for Industrial Spreader Stoker Boilers Fired with Wood and Bark Residue Fuels.* Oregon State University Press, Corvallis, OR.
33. Tillman, D., L. Reardon, M. Rollins, and E. Hughes. 1997. Cofiring Wood Waste for NOx Control in Cyclone Boilers: Identifying the Mechanisms. Proc. Third Biomass Conference of the Americas. Montreal, Canada. Aug 24-29. pp. 391-400.
34. Miller, S.F., B.G. Miller, and C.M. Jawdy. 2002. The Occurrence of Inorganic Elements in Various Biofuels and its Effect on Combustion

Behavior. Proc. 2002 International Joint Power Generation Conference. Phoenix, AZ. June 24 – 27.

35. Unterberger, S., C. Lopez, and K.R.G. Hein. 2001. EU-Project "Prediction of Ash and Deposit Formation for Biomass Pulverized Fuel Co-Combustion. Proc. Power Production in the 21st Century: Impacts of Fuel Quality and Operations. United Engineering Foundation Advanced Combustion Engineering Research Center. Snowbird, UT. Oct 28 – Nov 2.
36. Yan, L., R. Jensen, J. Laumb, and S. Benson. 2001. Predicting Ash Particle-Size and Composition Distribution from Coal-Biomass Cofiring. Proc. Power Production in the 21st Century: Impacts of Fuel Quality and Operations. United Engineering Foundation Advanced Combustion Engineering Research Center. Snowbird, UT. Oct 28 – Nov 2.
37. Korbee, R., J. Kiel, M. Zevenhoven, B. Skrifvars, P. Jensen, and F. Frandsen. 2001. Investigation of Biomass Inorganic Matter by Advanced Fuel Analysis and Conversion Experiments. Proc. Power Production in the 21st Century: Impacts of Fuel Quality and Operations. United Engineering Foundation Advanced Combustion Engineering Research Center. Snowbird, UT. Oct 28 – Nov 2.
38. Skrifvars, B., T. Lauren, R. Korbee, and P. Ljung. 2001. Ash Behavior in a Pulverized Wood Fired Boiler – a Case Study. Proc. Power Production in the 21st Century: Impacts of Fuel Quality and Operations. United Engineering Foundation Advanced Combustion Engineering Research Center. Snowbird, UT. Oct 28 – Nov 2.
39. Folkedahl, B., C. Zygarlicke, P. Hutton, and D. McCollor. 2001. Biomass for Energy – Characterization and Combustion Ash Behavior. Proc. Power Production in the 21st Century: Impacts of Fuel Quality and Operations. United Engineering Foundation Advanced Combustion Engineering Research Center. Snowbird, UT. Oct 28 – Nov 2.
40. Miles, T.R., T.R. Miles Jr., L.L. Baxter, B.M. Jenkins, and L.L. Oden. 1993. Alkali Slagging Problems with Biomass Fuels. Proc. First Biomass Conference of the Americas. Burlington, VT. Aug 30 – Sep 2. pp. 406 – 421.
41. Baxter, L.L. et. al. 1996b. The Behavior of Inorganic Material in Biomass-Fired Power Boilers—Field and Laboratory Experiences: Vol II of Alkali Deposits Found in Biomass Power Plants. SAND96-8225 Volume 2 and NREL/TP-433-8142.

42. Baxter, L.L. et. al. 1996a. The Behavior of Inorganic Material in Biomass-Fired Power Boilers: Field and Laboratory Experiences. Proc. Biomass Usage for Utility and Industrial Power. Engineering Foundation Conference. Snowbird, UT. April 28-May 3.
43. Baxter, L.L. 2000. Ash Deposit Formation and Deposit Properties: Final Report. SAND2000-8253. Sandia National Laboratories. Livermore, CA.
44. Tillman, D.A. 1994. Trace Metals in Combustion Systems. Academic Press, San Diego.
45. Tillman, D. and K. Hood. 1990. Incineration of Pulp Mill Sludge in High Efficiency Power Boilers. AIChE Annual Meeting, San Diego, CA. Aug 15 – 17.
46. Envirosphere Co. 1984. Final Biomass Ash Study. Prepared for the California Energy Commission. Contract No.: 500-81-037.
47. Greene, W.T. 1988. Wood Ash Disposal and Recycling Sourcebook. OMNI Environmental Services, Inc., Beaverton, OR.
48. Payette, K., T. Banfield, T. Nutter, and D. Tillman. 2002. Emissions Management at Albright Generating Station through Biomass Cofiring. Proc. 27th International Technical Conference on Coal Utilization and Fuel Systems. Coal Technology Association. Clearwater, FL. March 4-7. pp. 177-186.
49. Stultz, S.C. and J.B. Kitto (eds). 1992. Steam: Its Generation and Use, 40th Ed. Babcock & Wilcox, Barberton, OH.B
50. Tillman, D., A. Rossi, and W. Kitto. 1981. Wood Combustion: Principles, Processes, and Economics. Academic Press, New York.
51. Villesvik, Gordon and David A. Tillman. 1983. Cofiring of Dissimilar Solid Fuels: A Review of Some Fundamental and Design Considerations. American Power Conference, April 1983.
52. Irving, J. 1993. Wood Fired Power Plant Experience at the 50 MW McNeil Station. Proc. Strategic Benefits of Biomass and Waste Fuels. Washington, DC. March 30-April 1. EPRI, Palo Alto, CA.
53. Wiltsee, G.A. 1994. Biomass Energy Fundamentals, Vol 2: System Characteristics. Electric Power Research Institute, Palo Alto, CA.
54. Bergesen, C. and J. Crass (eds). 1996. Power Plant Equipment Directory, 2nd Ed. Utility Data Institute, a McGraw Hill Co., Washington, D.C.
55. Battista, J. and E. Hughes. 2001. Biomass Cofiring in the United States: A Survey of Past Tests. Proc. 26th International Technical

Conference on Coal Utilization and Fuel Systems. Clearwater, FL. March 5 – 8. pp. 227-236.

56. Boylan, D. 1993. Southern Company Tests of Wood/Coal Cofiring in Pulverized Coal Units. Proc. Strategic Benefits of Biomass and Waste Fuels. Washington, DC. March 30-April 1. EPRI, Palo Alto, CA.
57. King, J. 2002. Presentation Concerning the Renewable Program of JEA. Biomass Interest Group Meeting, Nov 7-8. Washington, D.C.
58. Tampa Electric Company. 2002. Biomass Test Burn Report: Polk Power Station Unit 1. Report to NETL-USDOE. April.
59. Raskin, N., J. Palonen, and J. Nieminen. 2001. Power boiler fuel augmentation with a biomass fired atmospheric circulating fluid-bed gasifier. Biomass and Bioenergy. 20(1): 471-481.
60. Tillman, D.A., K. Payette, T. Banfield, and S. Plasynski. 2002. Cofiring Demonstrations at Allegheny Energy Supply Co., LLC. Proc. 19th Annual International Pittsburgh Coal Conference. Pittsburgh, PA. Sep 23 – 26.
61. Tillman, D.A., K. Payette, and T. Banfield. 2003. Demonstrating Cofiring in PC and Cyclone Boilers at Allegheny Energy. Proc. Electric Power Conference, Houston, TX. March 3-6.
62. Battista, J.J., E.E. Hughes, and D.A. Tillman. 2000. Biomass cofiring at Seward Station. Biomass and Bioenergy. 19(6): 419-428.
63. Tillman, David A., Evan Hughes, and Sean. Plasynski. 1999. Commercializing Biomass-Coal Cofiring: The Process, Status, and Prospect. Proc. The Pittsburgh Coal Conference. Pittsburgh, PA. Oct 11-13.
64. Battista, J., D. Tillman, and E. Hughes. 1998. Clofiring wood waste with coal in a wall-fired boiler: initiating a 3-year demonstration program. Proc. Bioenergy '98. Madison, WI. Oct 4-8. pp. 243-250.
65. Laux, Stefan, John Grusha, and David Tillman. 2000. Co-firing of Biomass and Opportunity Fuels in Low NOx Burners. Proc. 25th International Technical Conference on Coal Utilization and Fuel Systems. ASME-FACT, USDOE, and Coal Technology Association. Clearwater, FL. March 6-9. pp. 571-582.
66. Tillman, D.A., M.L. Rollins, and L.D. Reardon. 1996. Cofiring Alternate Fuels in Coal-Fired Cyclone Boilers. Proc. American Flame Research Committee International Symposium. Baltimore, MD. Sep 30 – Oct 2.

67. Tillman, D. 1996. Cofiring Wood Waste in Utility Boilers: Results of Parametric Testing and Engineering Evaluations. Proc. 1996 International Joint Power Generation Conference. EC-Vol. 4. FACT-Vol. 21. ASME. Pp. 455-462
68. Tillman, D., R. Stahl, D. McLellan, D. Bradshaw, L. Reardon, M. Rollins, and E. Hughes. 1996. Fuel Blending and Switching for NO_x Control Using Biofuels with Coal in Cyclone Boilers. Proc. Biomass Usage for Utility and Industrial Power: an Engineering Foundation Conference. April 28-May 3. Snowbird, UT.
69. Tillman, David., James Campbell, Robert Stahl, Kristin Therkelsen, Christine Newell, and Evan Hughes. 1998. Cofiring Urban Wood Waste with Powder River Basin Coal at the Michigan City Generating Station. Proc. 23rd International Technical Conference on Coal Utilization and Fuel Systems. ASME-FACT, USDOE, and Coal Technology Association. Clearwater, FL. March 9-13. pp. 35-47.
70. Tillman, D., C. Newell, P. Hus, E. Hughes, and K. Therkelsen. 1998. Cofiring Biomass in Cyclone Boilers Using Powder River Basin Coal: Results of Testing at the Michigan City Generating Station. Proc. Bioenergy '98. Madison, WI. Pp. 285-294.
71. Hus, Patricia J. and David A. Tillman. 2000. Cofiring Multiple Opportunity Fuels with Coal at Bailly Generating Station. Biomass and Bioenergy. 19(6):385-394.
72. Tillman, David A. and Patricia Hus. 1999. Cofiring Multiple Opportunity Fuels for Cost-Effective Biomass Utilization. Proc. 4th Biomass Congress of the Americas. Oakland, CA. Aug 30 – Sep 2. pp.1349-1356.
73. Tillman, D.A., K. Payette, and J. Battista. 2000. Designer Opportunity Fuels for the Willow Island Generating Station of Allegheny Energy Supply Company, LLC. Proc. 17th Annual Pittsburgh Coal Conference. Pittsburgh, PA. Sep 11-14.
74. Tillman, D.A. and K. Payette. 2001. Developing a Designer Opportunity Fuel System for Willow Island Generating Station. Proc. 26th International Technical Conference on Coal Utilization and Fuel Systems. ASME-FACT, USDOE, and CSTA. Clearwater, FL. March 5-8.
75. Tillman, D., K. Payette, T. Banfield, and G. Holt. 2002. Firing Sawdust and Tire-Derived Fuel with Coal at Willow Island Generating Station. Proc. 27th International Technical Conference on Coal

Utilization and Fuel Systems. Coal Technology Association. Clearwater, FL. March 4-7. pp. 165-176.

76. Plasynski, Sean I., Raymond Costello, Evan Hughes, and David Tillman. 1999. Biomass Cofiring in Full-Sized Coal-Fired Boilers. . Proc. 24th International Technical Conference on Coal Utilization and Fuel Systems. ASME-FACT, USDOE, and Coal Technology Association. Clearwater, FL. March 8-11. pp.281-292.

CHAPTER 5: HERBACEOUS AND AGRICULTURAL BIOMASS OPPORTUNITY FUELS

5.1. Introduction

Herbaceous biomass fuels—agricultural and agribusiness wastes, herbaceous crops such as switchgrass, and the like—are fundamentally different from woody biomass fuels and must be treated separately. They are becoming potentially significant sources of energy in numerous countries from Denmark to China. Utilities such as Midkraft and Elkraft in Denmark routinely fire straw in electricity generating boilers and combined heat and power boilers [1]. Such practices are in response to mandates codified in Danish law. Further, herbaceous and agricultural energy sources such as crop wastes and vineyard wastes have been used substantially in such states as California, and they are being pursued as a potential long-term energy resource throughout the USA [2]. Utilities such as JEA, for example, have made investments in research designed to identify the most suitable herbaceous crops and to quantify the yields of such crops in energy terms [3].

China uses a substantial amount of herbaceous material for energy. Crop residues from the production of rice, corn, and wheat are currently used extensively for energy in this country; over 40 percent of the biomass energy generated in China comes from herbaceous and agricultural sources [4]. China consumes biomass at a rate equal to $260x10^6$ tonnes of bituminous coal annually [5], an amount equal to about 6.9 EJ ($6.5x10^{15}$ Btu), and the consumption of herbaceous materials for energy in this Asian country is on the order of 2.7 EJ ($2.6x10^{15}$ Btu).

Further, this consumption of herbaceous crops is projected to grow significantly within the Chinese economy in the next decade [4].

Numerous researchers have documented the fundamental differences between herbaceous biomass and woody biomass (see, for example, Hansen et. al. [1], Baxter et. al. [6], Tillman [7]). The herbaceous materials are significantly higher in ash than their woody counterparts; and this ash has significant slagging and fouling characteristics. Many of the herbaceous biofuels such as alfalfa residues can be very high in nitrogen [8]. Herbaceous crops, as fuel, can also contain significant concentrations of chlorine [8]. The herbaceous materials such as switchgrass can have very low bulk densities, and can present serious material handling and logistics issues as well [9, 10]. The consequence of all of these factors contributes to the conclusion that herbaceous and agricultural materials are potentially significant as energy sources, but they are a separate class of opportunity fuels.

5.1.1. Types of Herbaceous Biomass Fuels

Several types of herbaceous biomass fuels are used, or proposed, as opportunity fuels. These include crops grown specifically for fuel purposes as discussed historically by Szego, Kemp, and co-workers [11, 12] and more recently by researchers at the Oak Ridge National Laboratory (ORNL) and other institutions (see, for example, [13-16]). The most prominent of these is switchgrass, although Giant Reed has also been discussed [3] along with miscanthus and other crops. Switchgrass is the prominent crop chosen for evaluation by ORNL researchers and by other institutions.

Field crop residues also are significant. Field crop residues such as straw are commonly used as energy sources in Scandanavia [1, 17]; Denmark, particularly, focuses upon straw burning. Field crop residues such as alfalfa stalks not used in animal feed production have been considered in the USA along with corn stover, rice straw, and other residues from the growing and harvesting of grains [6, 18,19]. These residues have long been evaluated as fuels despite difficulties in gathering, concentrating, and transporting them.

Agribusiness residues also have been long considered in the herbaceous and agricultural fuels arena. These materials include a wide variety of potential fuels: bagasse from sugar cane processing, out-of-date corn seed, corn cobs, rice hulls, oat hulls, nut hulls (e.g., walnut hulls) and stone fruit pits (e.g., peach pits, olive pits), vineyard trimmings, orchard trimmings, tomato pummace from the manufacture of katsup, winery pummace from the squeezing of grapes, and a host of other locally available agribusiness wastes [19]. Bagasse has been the primary source of energy for the cane sugar industry [20]. Cotton gin trash has also been used for energy purposes in the US. [21].

5.1.2. Sources and Uses of Herbaceous Materials

Herbaceous and agricultural materials can be obtained from farms growing crops—either as foodstuffs or as fuel—or from a variety of agribusinesses. These a agribusinesses include food processors from the grain processing industry, from the animal food preparation industry, from fruit canners and packagers, vineyards, vintners, processed food manufacturers, and a host of other organizations.

In recent years there have been numerous efforts to gather farmers together into cooperatives and other organizations within the USA in order to produce sufficient material to support fuel applications. These organizations have promoted projects such as the proposed alfalfa stalk gasification efforts in Minnesota [22] and the ongoing efforts in Chariton Valley, IA to grow switchgrass for cofiring in the Ottumwa Generating Station of Alliant Energy [23]. Similar efforts are also underway to promote the use of switchgrass for cofiring in the southeast, principally at Southern Company generating stations [2, 7]

Since these materials have been produced for decades, producers of herbaceous residues have developed other markets for some of the materials. Such markets include animal bedding, paper and low-grade board manufacturing, and opportunity fuel applications. Field crop wastes are frequently plowed under or burned, depending upon local regulations, as a means of destruction. In developing countries field crop wastes are gathered for use in domestic fuel applications as well as small industrial applications.

5.1.3. Approach of this Chapter

It is impossible to provide detailed data on all herbaceous and agricultural material used or considered for energy purposes. The approach of this chapter, therefore, is to focus upon switchgrass for fuel characterization, making comparisons to other fuels as appropriate. Switchgrass is chosen because it is the apparent fuel of choice for energy crops as developed by USDOE. It is typically low in moisture and modest in heating value. Otherwise it shows characteristics common to most of the herbaceous materials. While switchgrass is presented as the primary herbaceous biomass, other materials that have been used or proposed as fuel are presented for comparative purposes.

Case studies reported in this chapter will include not only switchgrass cofiring applications, but also the use of straw for energy purposes. The chapter will extend its reach beyond switchgrass and the herbaceous materials to other agricultural fuels—principally manures used for energy purposes—when the primary discussion is complete.

5.2. Fuel Characteristics of Switchgrass and Related Agricultural Biomass Materials

Fuel characteristics of importance include fuel density, proximate and ultimate analysis, chemical structure and reactivity, and inorganic constituents and chemistry [24]. These characteristics are somewhat similar to—yet quite distinct from—the compositional aspects of woody biomass materials described previously in Chapter 4.

5.2.1. Density of Switchgrass and Related Materials

Switchgrass and related herbaceous materials are a very low density, modest heat content set of fuels. For the field crops such as switchgrass, bulk densities are less than 0.064 kg/m^3 (5 lb/ft^3) loose, and about 0.103 kg/m^3 (8 lb/ft^3) when baled [7]. These values compare to typical coal densities on the order of 0.666 kg/m^3 (52 lb/ft^3). The values for switchgrass are comparable to loose and baled straws, hay, and other field materials. Densities of vineyard prunings, stone fruit pits, and related products are higher—approaching the densities of woody biomass [18]. Table 5.1. presents typical bulk densities of switchgrass and other agricultural materials.

Table 5.1. Bulk Densities of Switchgrass and Other Agricultural Materials

Material	Bulk density (kg/m^3)	Bulk density (lb/ft^3)
Baled switchgrass	0.102	8.0
Cotton gin trash	0.078	6.1
Peach pits	0.384	30.0
Rice hulls	0.232	18.1
Walnut shells	0.426	33.3
Peanut shells	0.176	14.2
Vineyard prunings	0.170	13.3
Hybrid corn seed	0.320	25.0

Note that the densities of all but the switchgrass are for fuels produced at <2 mm particle sizes. Sources: [7, 18, 24].

Note that Table 5.1 shows a wide diversity of agricultural materials. All of these materials have been considered for fuel applications by industries, independent power producers (IPP's), and utilities in the USA and Europe, particularly in the UK. Many of these fuels have been used in small power plants within the state of California or the southeast, and related agricultural areas. Note, also, that most of the herbaceous and agricultural materials have densities that are significantly lower even than wood fuels.

Like the woody biofuels, the herbaceous biomass forms are porous solids, with significant fractional void volumes. Unlike the woody biofuels, however, the agricultural materials contain significant concentrations of inorganic matter, making the calculation of porosity less precise, and less prone to ready quantification by formula.

5.2.2. *Proximate and Ultimate Analysis of Switchgrass and Related Agricultural Materials*

The proximate and ultimate analyses for switchgrass and related herbaceous materials are shown in Table 5.2. Note that the moisture content for such materials is typically quite low relative to most biomass fuels.

Table 5.2. Proximate and Ultimate Analyses for Switchgrass and Selected Herbaceous Materials as Fuel

Parameter	Herbaceous Fuel			
	Fresh Switchgrass	Weathered Switchgrass	Reed Canary Grass	Mulch Hay
Moisture %	15	15	65.2	19.5
Proximate Analysis (wt % dry basis)				
Volatiles	76.18	81.8	76.1	77.6
Fixed Carbon	16.08	14.8	19.8	17.1
Ash	7.74	3.4	4.1	5.3
Ultimate Analysis (wt % dry basis)				
Carbon	46.73	49.4	45.8	46.5
Hydrogen	5.88	5.9	6.1	5.7
Nitrogen	0.54	0.4	1.0	1.7
Sulfur	0.13	0.3	0.1	0.2
Oxygen	38.99	40.6	42.9	40.6
Ash	7.74	3.4	4.1	5.3
Higher Heating Value (dry basis)				
MJ/kg	18.04	18.97	16.54	18.76
Btu/lb	7750	8150	7103	8058

Sources: [24-26].

Proximate and ultimate analysis data are also available for other agricultural biomass fuels such as rice hulls, vineyard prunings, and other related potential opportunity fuels. These are shown in Table 5.3. Note the variability in these fuels with respect to moisture and ash contents. Note, also, the variability in fuel nitrogen contents. The cotton gin trash, vineyard prunings, and rice hulls are particularly high in nitrogen content.

The higher heating values of the residues shown in Table 5.3, with the exception of out of date hybrid corn seed, are substantially lower than those associated with woody biomass or coal, and those associated with switchgrass grown for energy purposes. Much of this can be attributed to the higher ash contents associated with herbaceous materials—particularly residues from agribusinesses. The density and heat content of the hybrid corn seed has made it a desirable herbaceous opportunity fuel when conditions are favorable.

Table 5.3. Proximate and Ultimate Analyses for Selected Agricultural Materials as Fuel

Parameter	Herbaceous Fuel				
	Cotton Gin Trash	Rice Hulls	Vineyard Prunings	Hybrid Seed Corn	Bagasse
Moisture %	7-12	7-10	20-40	12	45
Proximate Analysis (dry wt %)					
Volatiles	75.4	63.6	74.9	78.32	86.62
Fixed Carbon	15.4	15.8	14.3	20.15	11.95
Ash	9.2	20.6	10.8*	1.53	2.44
Ultimate Analysis (dry wt %)					
Carbon	42.77	38.30	47.99	46.00	48.64
Hydrogen	5.08	4.36	5.65	6.13	5.87
Nitrogen	1.53	0.83	0.86	1.84	0.16
Sulfur	0.55	0.06	0.08	0.18	0.04
Oxygen	35.38	35.45	39.61	44.32	42.82
Ash	14.69	21.00	5.81	1.53	2.44
Higher Heating Value (dry basis)					
MJ/kg	15.60	14.90	16.81	19.14	19.01
Btu/lb	6700	6400	7220	8222	8165

Sources: [data from 18, 26]. Note*: high ash content due to embedded dirt in sample.

The proximate and ultimate analyses lead to derived ratios and values that relate specifically to combustion system performance, including measures of reactivity to be discussed subsequently, and measures of pollution potential. The measures of pollution potential include kg nitrogen and sulfur (or SO_2)/ GJ (or expressed in lb/10^6 Btu), and kg ash or inorganic matter/GJ of fuel. These measures of pollution potential are shown in Table 5.4. Note that many of the herbaceous fuels are not totally sulfur free, and most are high in nitrogen. The high nitrogen content is expected based upon the fertilization practices in the agricultural community. These practices also contribute to the inorganic matter compositions of herbaceous materials—particularly the

Table 5.4. Pollution Potential Measures from the Ultimate Analyses of Herbaceous Materials

Fuel	Pollution Potential Measure: kg/GJ (lb/10^6 Btu)		
	Fuel Nitrogen	Fuel Sulfur[1]	Fuel Ash
Fresh Switchgrass	0.300 (0.697)	0.072 (0.168)	4.294 (9.987)
Weathered Switchgrass	0.211 (0.491)	0.158 (0.368)	1.794 (4.172)
Reed Canary Grass	0.605 (1.408)	0.061 (0.141)	2.482 (5.772)
Mulch Hay	0.907 (2.110)	0.107 (0.248)	2.828 (6.577)
Cotton Gin Trash	0.982 (2.284)	0.353 (0.821)	9.428 (21.93)
Rice Hulls	0.558 (1.297)	0.040 (0.094)	14.11 (32.81)
Vineyard Prunings	0.512 (1.191)	0.048 (0.111)	3.460 (8.047)
Hybrid Corn Seed	0.962 (2.238)	0.094 (0.219)	0.800 (1.861)
Bagasse	0.086 (0.20)	0.017 (0.04)	0.55 (2.98)

Note: [1]O_2 values in kg/GJ or lb/10^6 Btu are double the fuel sulfur values.
Sources: [7, 19, 22-24, 26].

concentrations of potassium (K) and phosphorous (P) in the ash of herbaceous material.

The significant variability in the herbaceous and agricultural materials used as opportunity fuels makes their use potentially more challenging. Not shown in these data are concentrations of chlorine. However, chlorine concentrations can be significant and problematical in herbaceous and agricultural materials. Typical values for switchgrass are 0.08 – 0.16 percent (800 – 1600 mg/kg, or ppmw, dry basis) [6, 26] while rice straw can contain as much as 0.50 percent (5000 mg/kg) [26] and hybrid seed corn can contain 0.078 percent chlorine (780 mg/kg) [26, 27].

The chlorine in herbaceous biomass typically occurs as a chloride ion, resulting largely from local soil conditions [28]. This chlorine can present significant operational problems in terms of corrosion, as well as contributing to environmental concerns. The chlorine facilitates the mobility of many inorganic compounds, most notably potassium. Chlorine concentrations can govern the extent to which alkali metals that vaporize during the combustion process, and can result in the formation of highly corrosive potassium chloride (KCl) deposits on boiler tubes [27]. The concentration of chloride ions in the fuel is another significant indicator of slagging and fouling potential [27].

5.2.3. Reactivity of Switchgrass and the Herbaceous Materials

Fuel reactivity can be inferred from the proximate and ultimate analysis, using fixed carbon/volatile matter ratios or their inverse, and using hydrogen/carbon (H/C) and oxygen/carbon (O/C) atomic ratios. Reactivity can be measured more precisely, however, using drop tube reactor (DTR) and carbon 13 nuclear magnetic resonance (^{13}C NMR) techniques as discussed in Chapters 2 and 4. Using DTR techniques, reactivity can be measured in terms of maximum volatile yield, the volatile release patterns of fuel carbon and fuel nitrogen, and the kinetics of fuel devolatilization [29]. ^{13}C NMR analysis can be used to obtain structural information in support of the DTR measurements. Char reactivity—the kinetics of char oxidation—can then be measured by thermogravimetric analysis (TGA).

The proximate and ultimate analysis measures of fuel reactivity have been calculated for the range of herbaceous biomass fuels discussed previously in this chapter. The full suite of reactivity measurements has been performed for switchgrass, focusing upon both fresh and weathered switchgrass. These results are shown below.

5.2.3.1. Reactivity Measures Derived from the Proximate and Ultimate Analysis

Three direct measures of reactivity can be derived from the proximate and ultimate analyses shown in Tables 5.2 and 5.3: fixed carbon/volatile matter ratio or its inverse, hydrogen/carbon atomic ratio, and oxygen/carbon atomic ratio. These are shown in Table 5.5. Green [30] demonstrates that all fuels, including the biomass fuels, conform to the following volatility relationship:

$$V_T = 62([H]/6)([O]/25)^{½} \qquad [5\text{-}1]$$

Where V_T is total volatile matter, H is hydrogen weight percentage, and O is oxygen weight percentage. All calculations are on a dry, ash free, sulfur free, and nitrogen free (DASNF) basis. On a moisture and ash free basis (MAF), a related equation can be constructed as follows:

$$V_{Td} = 0.2252 + 1.5435(O_d) - 0.3709(O_d)^2 \qquad [5\text{-}2]$$

Table 5.5. Measures of Reactivity for Selected Herbaceous Biomass Fuels

Fuel	Measure of Reactivity			
	FC/VM[1]	VM/FC[2]	H/C[3]	O/C[4]
Fresh Switchgrass	0.21	4.74	1.50	0.63
Weathered Switchgrass	0.18	5.53	1.42	0.62
Reed Canary Grass	0.26	3.84	1.59	0.70
Mulch Hay	0.22	4.54	1.46	0.66
Cotton Gin Trash	0.20	4.90	1.42	0.62
Rice Hulls	0.25	4.03	1.36	0.69
Vineyard Prunings	0.19	5.24	1.40	0.62
Corn Seed	0.26	3.90	1.59	0.72
Bagasse	0.14	7.16	1.44	0.66

Notes: [1]Fixed carbon/volatile matter from proximate analysis
[2]Volatile matter/fixed carbon from proximate analysis
[3]Hydrogen/carbon atomic ratio from ultimate analysis
[4]Oxygen/carbon atomic ratio from ultimate analysis

Where V_{Td} is total volatile matter on a decimal basis (rather than a percentage basis) and O_d is the weight of oxygen in the fuel on a decimal basis. The r^2 for this equation is 0.953 over a broad range of biomass and coal fuels. What is apparent from Table 5.5 is that the herbaceous biofuels all exhibit highly similar reactivities—and that these reactivities are comparable to the reactivities associated with woody biofuels, where typical FC/VM ratios are on the order of 0.24, and where H/C and O/C ratios are typically 1.45 and 0.66 respectively.

5.2.3.2. Drop Tube Reactor and Carbon 13 Nuclear Magnetic Resonance (^{13}C NMR) Measurements of Reactivity

Drop tube reactor testing has been performed by The Energy Institute of Pennsylvania State University on the fresh and weathered switchgrass to document its reactivity constants and its maximum volatile yield, using the methodology previously described in Chapter 2. Char reactivity was determined by Thermogravimetric analysis. These measures of reactivity were further supported by ^{13}C NMR measurements of aromaticity, average number of aromatic carbons per cluster, and concentration of carbon atoms in methoxy functional groups [7,25, 31]. Figures 5.1 and 5.2 present the devolatilization reactivities of fresh and weathered switchgrass as Arrhenius equations; note that the temperatures used to determine these kinetic equations were drop tube reactor temperatures, rather than particle temperatures. Figure 5.3 presents the maximum volatile yield from these fuels, and compares these volatile yields to Pittsburgh seam coal and Black Thunder PRB coal.

Note that these fuels are highly reactive, with low activation energies and with high maximum volatile yields. Further, when being tested in the DTR at temperatures associated with typical combustion conditions in large boilers, maximum volatile yield for the switchgrass is on the order of 90+ percent. Baxter [8] has reported maximum volatile yields for switchgrass as low as 80 percent when measured on a comparable basis.

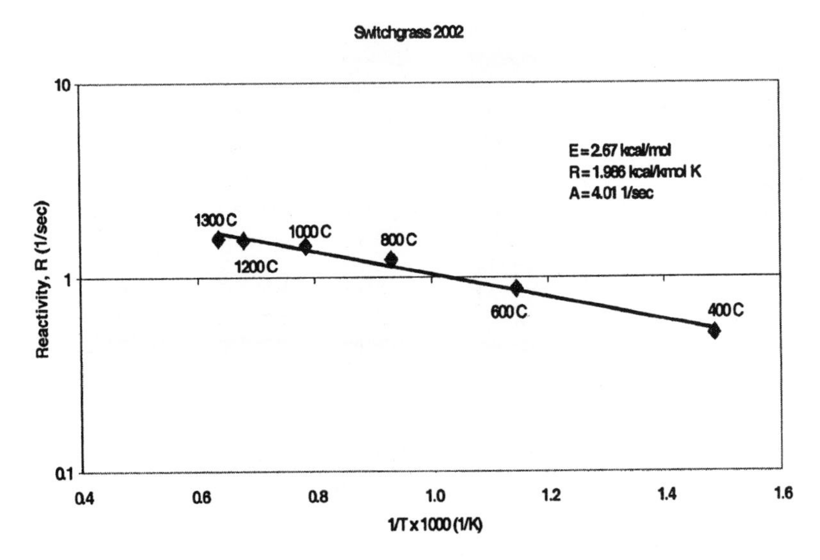

Figure 5.1. Devolatilization reactivity of fresh switchgrass
Source: [7, 9, 29]

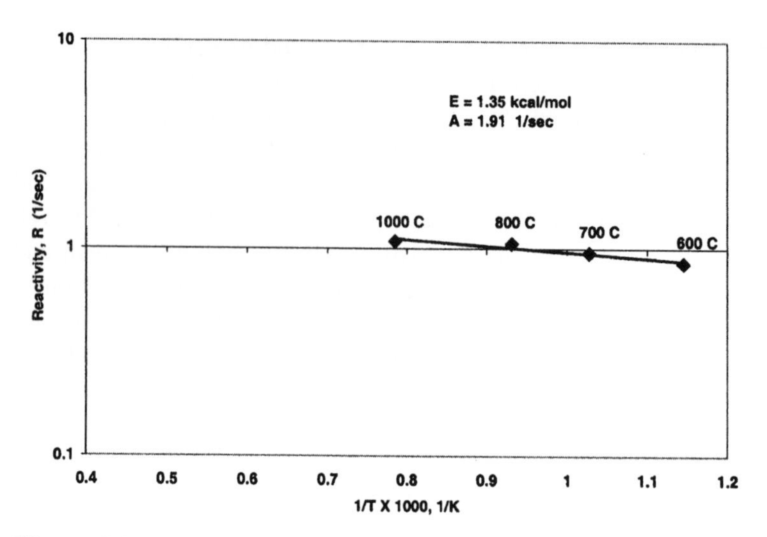

Figure 5.2. Devolatilization reactivity of weathered switchgrass
Source: [7, 9, 29]

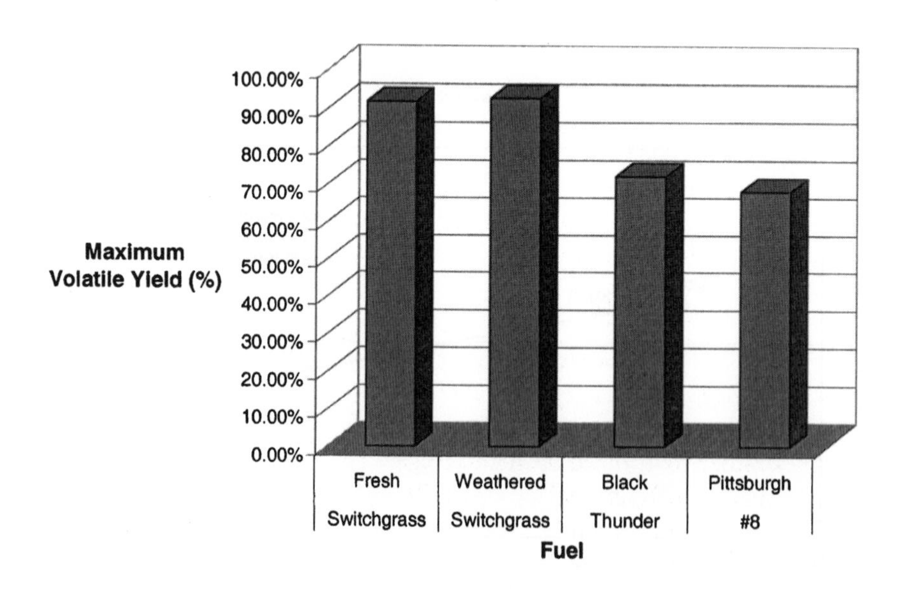

Figure 5.3. Maximum volatile yield of fresh and weathered switchgrass compared to reference coals
Source: [7, 9, 29]

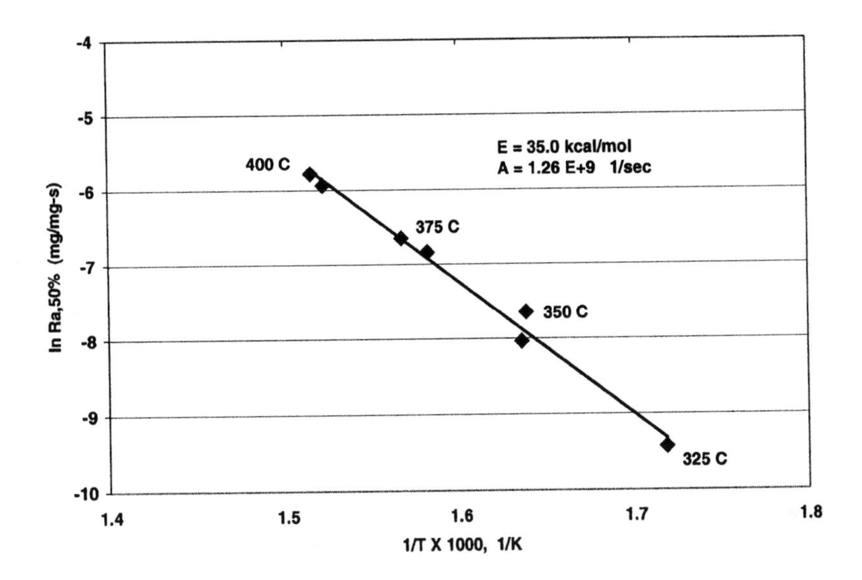

Figure 5.4. Char Oxidation Kinetics for Switchgrass
Source: [7, 9, 29]

The char is also quite reactive as is shown in Figure 5.4, the kinetics of switchgrass char oxidation. The activation energy associated with switchgrass char oxidation is considerably lower than that associated with any of the coals. This reactivity is pronounced at lower temperatures. Because the pre-exponential constant or frequency factor is also low, switchgrass char reactivity can be lower than that for some coals at the high temperatures.

The ^{13}C NMR characterization was applied only to the weathered switchgrass [7, 9]. It showed an aromaticity of 10 percent—just slightly higher than the aromaticity of sawdust. Like sawdust, the average number of aromatic carbons/cluster was 6, indicating single aromatic structures rather than fused aromatic rings. The concentration of carbon atoms in methoxy functional groups ($-OCH_3$) for weathered switchgrass was 7.9 percent—just slightly lower than the 8.7 percent of carbon atoms in $-OCH_3$ functional groups found in sawdust.

Nitrogen reactivity is also of significant concern in the use of herbaceous materials. As with the kinetics, switchgrass is used to represent the herbaceous biomass class of opportunity fuels. As discussed in prior chapters, the extent to which the nitrogen is released as volatile matter during the devolatilization processes of combustion governs the ability of any unit to control NO_x emissions with staged combustion and other combustion processes. Further, the volatile evolution pattern of the nitrogen, relative to carbon (representing the total mass of fuel), plays a significant role in the ability of a unit to use staged combustion and other combustion techniques for NO_x control.

Some 92 percent of the nitrogen in fresh switchgrass is released as volatile matter during combustion, while 85 percent of the nitrogen in weathered switchgrass is released as volatile matter during the devolatilization process of combustion [7]. The volatile nitrogen evolution patterns of fresh and weathered switchgrass vary dramatically, however, as is shown in Figures 5.5 and 5.6. With fresh switchgrass, the volatile nitrogen release lags behind the volatile carbon release, but only slightly. However there is a dramatic lag between volatile nitrogen release and volatile carbon release with weathered switchgrass.

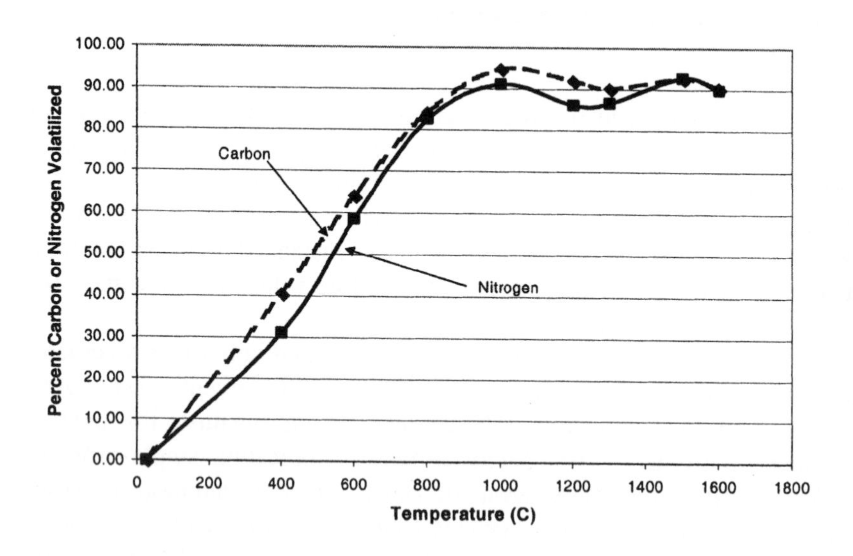

Figure 5.5. Volatile nitrogen and carbon release from fresh switchgrass as a function of temperature

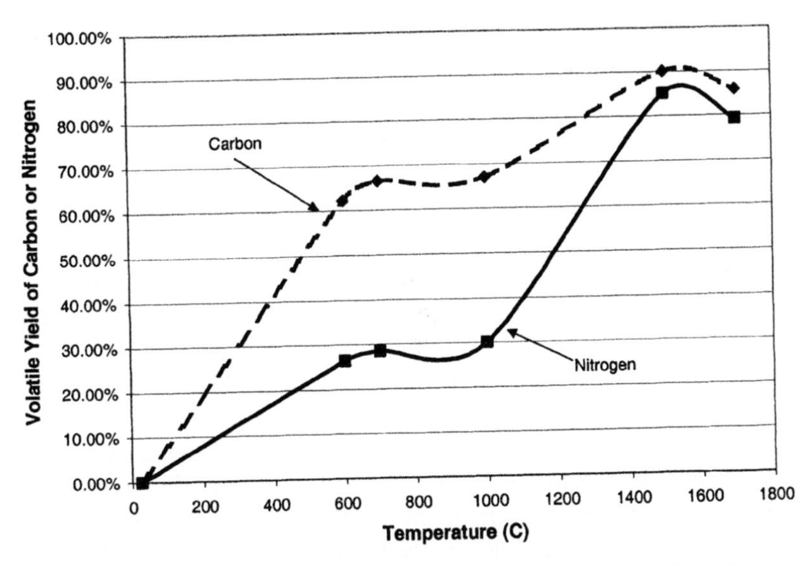

Figure 5.6. Volatile nitrogen and carbon evolution as a function of temperature for weathered switchgrass

The change in volatile nitrogen release as a function of temperature for weathered switchgrass, compared to fresh switchgrass, is consistent with much of the research on this fuel. Wiselogel et. al. [32] have shown that storage of round bales of switchgrass where the biomass is subject to rainfall and weathering does change the composition—particularly in the outer layer of the bale. On a total bale basis their research shows a consistent increase in nitrogen content. Further, and more significantly, their research shows a consistent loss of volatile matter as a function of weathering—particularly the most volatile non-structural components of the switchgrass.

The nitrogen evolution patterns for switchgrass are different from those associated with sawdust and, to a lesser extent, urban wood waste. However, Baxter [8] has reported that Alfalfa stalks, which are inherently high in nitrogen [29], have a nitrogen release pattern that is closer in nature to sawdust. The nitrogen volatiles leave preferentially to the carbon volatiles, even at low temperatures.

The herbaceous biomass fuels, then, are highly reactive. This is consistent with their overall chemical composition as shown above.

However, as herbaceous crops such as switchgrass, hay, and other products are stored in the weather in order to ensure a year-round fuel supply, the reactivity changes. This change is particularly pronounced with respect to volatile nitrogen evolution.

5.2.4. Ash Chemistry for Herbaceous Biomass Fuels

The ash chemistry associated with herbaceous biomass is particularly significant when considering these opportunity fuels. Research in the USA [6, 24, 33-35] and in Europe [1, 17] has demonstrated that these herbaceous fuel materials—high in ash content and high in alkalinity—present unique issues associated with utilization of these opportunity fuels. Ash deposition has been evaluated extensively in Europe as a consequence of programs designed to fire herbaceous biomass fuels—particularly straws, miscanthus, and related materials [1, 36-39]. When cofired in fluidized beds, herbaceous biomass fuels can cause bed agglomeration, hot spots, and potentially defluidization; in stokers such materials can create large deposits impeding effective air distribution through the grate while in pulverized coal boilers such materials can cause excessive slagging and fouling [38].

Traditional ash analyses result in ash elemental analyses such as those shown in Table 5.6. Miles et. al. [33] proposed a rule of thumb for evaluating the slagging and fouling potential of biomass fuels based upon ash elemental analyses as shown above. Fuels with alkali (Na_2O and K_2O) concentrations below 0.172 kg/GJ (0.4 lb/10^6 Btu) have only a slight opportunity to experience slagging and fouling, while fuels with alkali concentrations >0.344 kg/GJ (0.8 lb/10^6 Btu) are certain to experience significant slagging and fouling. Table 5.7 provides such indices for selected herbaceous biomass materials.

A more precise and definitive approach to biomass ash characterization has been developed and employed by researchers at The Energy Institute of Pennsylvania State University [34 – 35], Sandia National Laboratories [6], and numerous commercial and industrial laboratories. This procedure, previously described in Chapter 4, involves successive leaching of the fuel sample in distilled water, ammonium acetate, and hydrochloric acid. The most reactive materials leach in distilled water and ammonium acetate. These include salts of potassium and sodium which readily vaporize when subjected to high temperatures

Table 5.6. Ash Analyses of Switchgrass and Other Herbaceous Crops

Parameter	Fuel				
	Fresh Switchgrass	Weathered Switchgrass	Alfalfa Stems	Wheat Straw	Rice Straw
SiO_2	65.18	65.42	1.44	55.70	73.00
Al_2O_3	4.51	6.98	0.60	1.80	1.40
TiO_2	0.24	0.34	0.05	0.00	0.00
Fe_2O_3	2.03	3.56	0.25	0.70	0.60
CaO	5.60	7.14	12.90	2.60	1.90
MgO	3.00	3.17	4.24	2.40	1.80
Na_2O	0.58	1.03	0.61	0.90	0.40
K_2O	11.60	7.00	40.53	22.80	13.50
P_2O_5	4.50	2.80	7.67	1.20	1.40
SO_3	0.44	2.00	1.60	1.70	0.70
CO_2	0.00	0.00	17.44	0.00	0.00
Base/Acid Ratio	0.33	0.30	28.00	0.51	0.24

Sources: [6, 29]

Table 5.7. Slagging and Fouling Index for Selected Herbaceous Biomass Fuels

Fuel	Miles Slagging and Fouling Index kg/GJ (lb/10^6 Btu) of K_2O and Na_2O
Wheat Straw	1.33 (3.10)
Sunflower Hulls	2.51 (5.83)
Alfalfa Stems	1.45 – 1.97 (3.38 – 4.59)
Fresh Switchgrass	0.34 – 0.52 (0.80 – 1.22)
Weathered Switchgrass	0.14 – 0.22 (0.33 – 0.51)
Reed Canary Grass	0.51 (1.18)
Rice Straw	0.78 (1.82)
Bagasse (*)	0.064 (0.148)

Note: Bagasse is a washed product; all of the most reactive ash elements are leached out in the sugar cane processing activities.

Sources: [8, 22, 29, 33, 34]

associated with combustion systems [36]. These also include highly reactive calcium compounds as well. Carbonates and sulfates are commonly found in the acid soluble fraction, while silicates and sulfides—quite unreactive materials—are typically found in the residual. The chemical fractionation can then be enhanced by a number of other techniques: scanning electron microscopy (SEM) or computer controlled scanning electron microscopy (CCSEM), x-ray fluorescence (XRF), atomic absorption spectroscopy (AAS), and a variety of other techniques.

Thermodynamic modeling using the FactSage techniques provides additional insights into the phase changes of inorganic matter, and the consequent deposition characteristics of ash materials [34]. Of these techniques, chemical fractionation has shown the most immediate benefit in analyzing the inorganic fraction of biomass fuels—and low rank coals.

Table 5.8 provides detailed chemical fractionation data for three herbaceous biomass fuels. Note the high reactivity associated with each biomass fuel. Table 5.9 provides detailed chemical fractionation data for the potassium content of four switchgrass samples, demonstrating the consistent reactivity of the inorganic constituents in herbaceous biomass. This reactivity is comparable to woody biomass, however the concentrations of inorganic material in the herbaceous biomass fuels are considerably higher than those for woody biomass fuels (e.g., 4.3 – 12.9 kg/GJ or 10 – 30 lb/10^6 Btu in herbaceous biomass fuels vs <0.4 – 3.0 kg/GJ or <1 – 7.1 lb/10^6 Btu for woody biomass fuels including urban wood waste). The consequence of the combination of reactivity and higher ash concentration implies that herbaceous biomass fuels present more potential slagging and fouling problems than woody biomass fuels.

FactSage modeling performed by The Energy Institute of Pennsylvania State University demonstrated that the addition of herbaceous biomass to coal—in typical cofiring blends of 10 – 20 percent (mass basis) did not necessarily significantly increase the percentage of inorganic matter entering the liquid phase; however it did decrease the viscosity of the liquid phase material [34]. Further, it did decrease the T_{250} temperature from a calculated 1452°C (2647°F) for 100 percent bituminous coal to 1243°C (2269°F) when firing a blend of 10 percent switchgrass/90 percent eastern bituminous coal, and 1247°C (2,276°F) when firing a blend of 20 percent switchgrass/80 percent eastern bituminous coal [34]. Further additions of switchgrass increase, rather than decrease, the T_{250} temperature.

Table 5.8. Chemical Fractionation Analyses of Herbaceous Biomass Fuels

Ash Constituent	Herbaceous Material		
	Reed Canary Grass	Switchgrass	Wheat Straw
Potassium			
Water soluble and ion exchangeable	97	78	89
Acid soluble	2	15	7
Insoluble	1	7	4
Sodium			
Water soluble and ion exchangeable	98	52	52
Acid soluble	0		0
Insoluble	2	48	48
Calcium			
Water soluble and ion exchangeable	92	82	39
Acid soluble	7	10	17
Insoluble	1	8	45
Magnesium			
Water soluble and ion exchangeable	95	80	70
Acid soluble	4	7	5
Insoluble	1	13	25
Aluminum			
Water soluble and ion exchangeable	25	0	8
Acid soluble	3	0	0
Insoluble	72	100	92
Silicon			
Water soluble and ion exchangeable	55	0	7
Acid soluble	0	0	0
Insoluble	45	100	93

Sources: [6, 29, 34, 35]

Table 5.9. Chemical Fractionation of Potassium in Switchgrass for 4 Samples

Parameter	Switchgrass Sample			
	A	B	C1	C2
Water Soluble	52	68	70	72
Ion Exchangeable	35	24	20	16
Acid Soluble	9	0	0	2
Residual	4	8	10	10

Source: [6]

The FactSage modeling highlights the complexity of interactions among the inorganic constituents of herbaceous biomass—and the interactions between the inorganic matter of the biomass and the base coal used in cofiring situations. Of particular significance are concentrations of silicon, potassium, sulfur, and chlorine. As previously noted, the chlorine can mobilize the potassium, and other alkali metals and alkali earth elements [29]. Chlorine, as potassium chloride, can then cause a low melting ash or slag. Potassium chloride is reasonably stable in the gas phase, but readily deposits on heat transfer surfaces within the boiler. The sulfur from coal, also in the gas phase during combustion processes, can then substitute for the chloride in alkali chloride deposits such as potassium chloride deposits. This substitution results in potassium sulfate deposits which are corrosive, but less corrosive than potassium chloride deposits [6, 8].

The silicon in the ash also contributes to this interaction. The high silicon concentration in the switchgrass ash probably contributes significantly to the increased ash/slag viscosities in higher percentage switchgrass/coal blends.

5.2.5. *Trace Metal Concentrations in Herbaceous Biomass*

Trace metal concentrations in herbaceous biomass materials are the final area of concern, particularly with the emphasis on mercury emissions management and the possible concern for selective catalytic reduction catalyst deactivation or poisoning as a result of cofiring straws and other herbaceous biomass fuels [1, 17]. The database for herbaceous materials is not extensive. Table 5.10 presents a general range of values.

Table 5.10. Trace Metal Concentrations in the Ash from Agricultural Materials (mg/kg)

Metal	Minimum	Maximum
Antimony	10	10
Arsenic	3.4	12
Barium	41	220
Beryllium	0.01	0.06
Cadmium	0.36	1.1
Chromium	11	20
Cobalt	2.9	14
Copper	14	31
Lead	21	55
Mercury	BDL	BDL
Nickel	4.4	5.8
Selenium	BDL	BDL
Vanadium	11	20
Zinc	40	190

Sources: [40 - 42]

Trace metals in herbaceous crops are typically a consequence of fertilizer practices, and economic activities in the area. Commercial fertilizers can contain as much as 4,570 mg/kg Cr, 430 mg/kg Pb, and 3,000 mg/kg Zn [42, 43]. Further, the presence of a large source of airborne metals (e.g., a metals smelter) can cause elevated levels of trace metals in herbaceous biomass. Significantly it is the arsenic concentration that can cause poisoning of the SCR catalyst, along with the possibility of catalyst blinding from the alkali metals. The arsenic concentrations in herbaceous biomass ash are typically lower than those associated with western low rank coals and lignites [42].

5.3. Case Studies in the Use of Herbaceous Biomass as an Opportunity Fuel

Given the characteristics of herbaceous biomass as an opportunity fuel, it is important to review some of the full-scale experience burning this material in large-scale utility boilers.

5.3.1. Cofiring Straw at the Studstrup Generating Station of Midkraft

Firing of straws became legislatively mandated in Denmark during the early 1990's, and consequently approaches were taken either to fire the material as the sole fuel on stoker grates, or to cofire this material with coal. Elkraft, a Danish utility, developed firing the material on grates in stoker-fired boilers. Midkraft and Elsamprojekt developed cofiring approaches at Grena Power Station using a fluidized bed boiler, and at Studstrup, where cofiring occurred in a wall-fired PC boiler. These projects are described by Weick-Hansen [1, 17], and Overgaard [44].

In the Studstrup demonstration, straw was received, prepared, and fired in a 150 MWe wall-fired PC boiler. The system was designed and installed by ELSAM/MIDTKRAFT in 1995 at the Studstrup Generating Station. The straw management system included an automatic fuel receiving system that received Hesston bales, measured their moisture automatically, and then stacked them in storage. The maximum moisture content for the rectangular Hesston bales was a nominal 15 percent.

The bales were stored in a concrete warehouse, and each bale was identified with respect to moisture content. Computer systems managed the storage, including the location of each bale acquired by the cofiring system. Bales were reclaimed automatically, using the computer records of moisture content, and transported to a debaler and then an initial shredder. The debaler involved removing the cords from the bales automatically, prior to feeding the bale to the initial shredder. From the initial shredder, the straw was conveyed to an air classification system for removal of rocks and tramp materials, and then to a hammer mill for secondary particle size reduction. After the straw underwent secondary particle size reduction it was immediately transported to the boiler for combustion with coal. Once the straw was moving, it was kept moving.

The boiler itself (now decommissioned) was equipped with 12 burners on 3 rows. The burners in the middle row were modified to transport straw particles down the center of the burner, with coal flowing outside the straw particles in an arrangement similar to that demonstrated at Seward Generating Station for cofiring of sawdust.

The Studstrup testing was highly successful and provided key insights into automating cofiring materials handling systems. In particular it demonstrated the extensive investment required for automated handling, and the desirability of 2-stage grinding for herbaceous materials whenever possible. It identified the need for rectangular bales, rather than round bales—a need further proven by the Ottumwa, IA cofiring demonstration of switchgrass. It identified the problems associated with debaling, and difficulties associated with cord removal from bales. As a consequence it demonstrated the high capital costs associated with cofiring herbaceous materials.

The Studstrup testing also provided initial insights into the potential for herbaceous biomass to cause contamination of catalysts employed in selective catalytic reduction (SCR) systems. Their initial testing, although limited in scope and in applicability, provided clear indication that significant research must be done in this area.

5.3.2. *Cofiring at Plant Gadsden*

The testing and demonstration at Plant Gadsden along with related testing in the Southern Research Institute (SRI) combustion facility as described by Bush et. al. [45] Zemo et. al. [46], Boylan et. al. [47] and summarized by Tillman [7] provided key additional insights into potential alternative designs for cofiring systems, along with cofiring results. This program was aimed at cofiring switchgrass in a 70 MW_e tangentially-fired boiler. The testing was designed to address efficiency and environmental impacts of cofiring, particularly focusing upon airborne emissions. Further, it was supported by extensive pilot scale testing by Southern Research Institute, providing key insights into the transition from pilot scale to commercial scale operations.

Much of the research at the Southern Research Institute/Southern Company program was devoted to fuel growing and fuel characterization. While much of the fuel growing/fuel characterization program is outside the scope of this discussion, the influence of various conditions on heat content and ash fusion temperature is instructive. Hand cut samples were analyzed for these parameters. Table 5.11 summarizes the influence of switchgrass type and harvesting regime on heat content and ash fusion temperature.

Testing of different soil types on Alamo switchgrass using hand cut samples show that heat content variations between clay, sandy loam, and sandy soil ranged from 18.75 MJ/kg to 18.97 MJ/kg (8082 Btu/lb to 8175 Btu/lb) on a dry basis; ash fusion temperatures ranged from 1112°C to 1138°C (2034°F to 2082°F). Fertilization of Alamo switchgrass from 0

Table 5.11. Influence of Switchgrass Variety and Harvesting Regime on Heat Content and Ash Fusion Temperature

Switchgrass Variety	Number of cuts per year	Ash content (%)	Heat content (Btu/lb, dry)	Ash fusion temperature (°F)
Alamo	1	3.08	8156	2061
Alamo	2	4.03	8130	2118
Cave-in-Rock	1	3.20	8128	2136
Cave-in-Rock	2	3.86	8138	2141
Kanlow	1	3.37	8189	2144
Kanlow	2	3.78	8140	2264

Source: [45]

lb N/acre to 200 lb/acre increased the heat content from 18.57 MJ/kg (7999 Btu/lb) to 18.88 MJ/kg (8135 Btu/lb) on a dry basis. Ash fusion temperatures increased from 1006°C (1844°F) to 1172°C (2142°F) [45].

The initial testing of switchgrass cofiring occurred in the SRI combustion facility. This facility is an up-fired furnace designed with a 1.75 MW_{th} ($6x10^6$ Btu/hr) capacity. The furnace itself is 8.62 m (28 ft) high, and has an internal diameter of 1.08 m (3.5 ft). It is designed for a gas residence time of 1.3 – 2.5 sec and a FEGT of about 1204°C (2200°F) [45]. The burner is mounted coaxially at the bottom of the furnace, and can impart swirl to simulate tangential firing to represent the combustion at Gadsden Station [45]. Cofiring testing at the SRI reactor involved blending pulverized switchgrass with Pratt steam coal. This cofiring testing demonstrated that cofiring of switchgrass could reduce emissions (see Figure 5.7). Note that there is a step change associated with cofiring when compared to firing coal alone.

The initial testing at SRI also documented materials handling difficulties associated with firing switchgrass. However the initial testing was sufficiently successful to support the full scale testing at Plant Gadsden.

The design of the system for Plant Gadsden was significantly different from the design for the Studstrup demonstration. The switchgrass bales were traditional round bales from farming operations; they were ground to final particle size in a single step, in a tub grinder

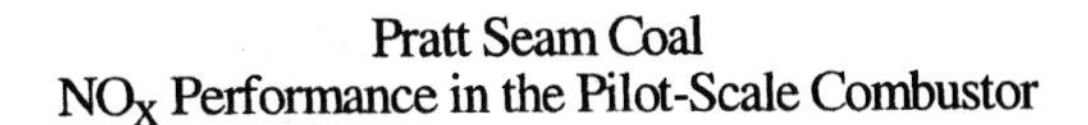

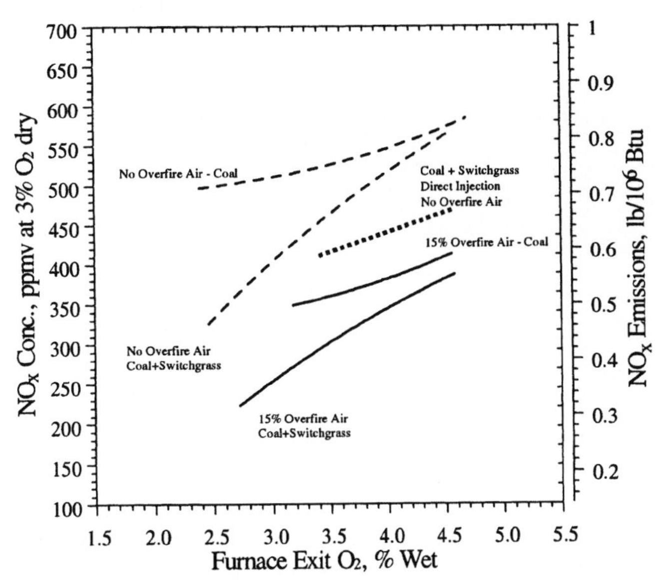

Figure 5.7. NOx Emissions Measured During Switchgrass Cofiring Testing at the SRI Combustor
Source: [45]

powered by a farm tractor. The ground switchgrass was conveyed to a small live-bottom surge bin and then picked up by an exhauster fan using an eductor design, and transported to the plant in a common duct. A splitter at the end of the duct directed the switchgrass to the two opposite corner injection penetrations of the boiler, where the switchgrass was fired.

The system was simple. Round switchgrass bales stocked out in one portion of the fuel yard, were delivered to the tub grinder where the switchgrass was ground to about 12.7 x 0 mm (½" x 0") to 19 x 0 mm (¾" x 0") particles as is shown in Figure 5.8. The ground switchgrass was then mechanically transported to a live bottom bin that metered the rate of switchgrass discharged to the boiler. The grass particles were transported about 154 m (500 ft) through a 0.3 m (12") dia duct. At the

Figure 5.8. Feeding a Round Bale to the Tub Grinder at Plant Gadsden

left front corner of the furnace, the duct was split into a pair of 0.2 m (8") ducts transporting the biomass to the burner inserts. The burner inserts, supplied by Foster Wheeler were full-throated burners that were adjustable to allow the tilt of the switchgrass injector to follow the tilt of the coal nozzles.

The splitter was a key element in the design. Flow through the splitter was not completely balanced, with the front left corner receiving 57 percent of the air flow and the rear right corner receiving 43 percent of the air flow. The switchgrass control system was largely manual. The main panel adjacent to the control room contained the Allen Bradley PLC that controlled startup and shutdown procedures, and monitored the system pressure for pipe plugs and system trips.

The actual testing was conducted using a synfuel as the base coal. Some 40 tests were conducted, with each test lasting 1 – 2 hours. In general the testing showed a boiler efficiency loss of 0.3 percent to 1.0 percent, depending upon the level of cofiring and boiler conditions. Efficiency losses associated with dry gas, moisture, and hydrogen accounted for this efficiency deterioration. At the same time SO_2, mercury, and fossil CO_2 emissions decreased with cofiring.

NO_x emissions did not decrease with the cofiring. Experimentation showed that the transport air potentially overcame the benefits of the biomass in NO_x reduction. The air/fuel ratio employed by the Plant Gadsden test exceeded 4:1 on a lb/lb basis. Recognizing that the tests were conducted with weathered switchgrass, there may have been fuel effects causing this failure to reduce NO_x emissions as well (see Chapter 2).

Since the initial tests, Southern Company has fired mixed grasses in this cofiring installation. Mixed grasses bale to a higher bulk density, and therefore are more difficult to grind. Two operational lessons were learned from the Plant Gadsden program of critical significance: 1) it is important not blend herbaceous materials with coal for pulverizing and storing in bunkers; the blend bridges very badly in the bunker and plugs the outlet (as was tested by Jenike and Johanson for Southern Company); and 2) it is important not store switchgrass in the surge bin overnight; it settles in and bridges, making the bin inoperable until the material has been removed. Once the switchgrass is moving, keep it moving.

The Gadsden station demonstration continues with mixed grasses. Included in this demonstration is the provision for testing SCR catalyst coupons for deactivation when cofiring herbaceous biomass with coal.

5.3.3. Cofiring switchgrass at the Ottumwa Generating Station

The largest single demonstration of switchgrass cofiring was the Chariton Valley project in Ottumwa, IA, where switchgrass cofiring has been tested in a 704 MW_e tangentially-fired boiler. The Ottumwa Generating Station in Chillicothe, IA fired 1,151 tonnes of switchgrass during the 2001-2002 winter, at rates up to 15 tonnes/hr. This represented up to 3 percent of the heat input to that unit. Again the results showed difficulties in materials handling; all of the round bales had to be rebaled into rectangular shapes in order to achieve a useful project. Cofiring was tested at this facility for a significant duration of time. The tests generally showed no conclusive emissions or efficiency impacts [10]. Table 5.12 compares the switchgrass cofired to the PRB coal used as the primary fuel at this station. Table 5.13 compares the average emissions produced when firing 100 percent coal, and when cofiring switchgrass with coal at this station.

In order to accomplish the cofiring of switchgrass at Ottumwa, a system was built to receive rectangular bales, debale them and produce

Table 5.12. Analysis of the Coal and Switchgrass Burned at Ottumwa Generating Station

Analytical Parameter	Fuel	
	Coal	Switchgrass
Proximate Analysis (wt % as received)		
Moisture	33.52	6.34
Ash	5.51	5.35
Volatile Matter	28.98	73.84
Fixed Carbon	32.00	14.48
Total	100.01	100.01
Higher Heating Value, MJ/kg (Btu/lb)	18.1 (7774)	17.4 (7458)
Ultimate Analysis (wt %, dry basis)		
Carbon	67.89	48.41
Hydrogen	4.48	5.06
Oxygen	17.58	40.16
Nitrogen	1.16	0.56
Sulfur	0.55	0.12
Chlorine	0.02	0.14
Ash	8.24	5.70
Total	99.92	100.15
Sodium Oxide (Na2O)	0.067	0.003
Potassium Oxide (K2O)	0.004	0.809

Sources: [7, 10]

particles suitable for firing, and transport them to the boiler. There the switchgrass was fired in one side of the dual furnace 704 MW_e unit. The cofiring demonstration at Ottumwa Generating Station has been taken

Table 5.13. Comparison of Average CEM Results for January 2001

Emissions Measure	Combustion Condition		
	Coal Only	Cofiring	Average of All Days
NO_x (ppmv)	176	187	182
SO_2 (ppmv)	247.8	245.7	246.7
Opacity (%)	16.58	16.41	16.48

Sources: [7, 10]

only to the first level projected by the project. A system is being designed for construction and operation that would increase the cofiring capability significantly. Ultimately this demonstration project will cofire switchgrass at about 5 percent of the heat input to the total boiler—essentially doubling the capability demonstrated in the initial testing.

The Ottumwa demonstration highlights the fact that cofiring of herbaceous materials can be accomplished in very large, modern boilers as well as smaller and older generating units. It demonstrates that cofiring can be achieved in such units without causing undue difficulties. It also demonstrates that cofiring of herbaceous biomass does not always achieve the emissions reductions that are often projected for such projects.

5.3.4. Other herbaceous biomass opportunity fuel demonstrations

Herbaceous biomass has long been fired in stoker boilers and fluidized bed boilers in California, where the agricultural community was encouraged by the California Energy Commission (CEC) to enter the energy arena. The CEC has had biomass programs for the agricultural community since the early 1980's.

Table 5.14 identifies many of these California projects. The firing of these materials in California led to the significant slagging and fouling problems associated with potassium and sodium contents, combined with chlorine contents, as investigated extensively by Miles et. al. [6] and Baxter et. al. [28] and discussed previously in this chapter. Wheat straws have been fired in fluidized bed boilers such as the Grena Power Plant of Midkraft since the mid 1990's.

Switchgrass has been cofired—under test conditions—at Blount St. Station of Madison Gas & Electric in Madison, WI [48]. At this 55 MW_e wall-fired boiler, cofiring of switchgrass at 10 percent on a heat input basis achieved significant SO_2 and NO_x reductions, with only modest efficiency losses. Blount St. Station is somewhat unique, in that it is equipped for cofiring waste paper and plastics, and includes a dump grate to capture and burnout larger fuel particles not entrained in the gaseous combustion products. Cofiring at Blount St. Station also demonstrated the potential high costs of switchgrass as a fuel (e.g., $61/ton) due to costs associated with harvest and transport of this low density material.

Table 5.14. Herbaceous Biomass Burning Boilers in California

Owner/Operator	Location	Boiler type	Capacity (MWe)	Fuel
Delano Energy Co	Delano, CA	Bubbling fluidized bed	27	Agricultural prunings, urban wood fuel
CMS Generation	Imperial	Stoker-fired grate	20	Straw, urban wood fuel
Mendota Biomass Power, Ltd	Mendota	Circulating fluidized bed	28	Nut hulls, nut shells, agricultural wastes, wood
Sithe Energies, Inc.	Marysville	Circulating fluidized bed	18	Agricultural wastes, urban wood fuel
Wheelabrator	Shasta	Stoker-fired grate	55	Wood, non-recyclable paper
Woodland Biomass Power, Ltd	Woodland	Circulating fluidized bed	28	Agricultural wastes, urban wood fuel

Source: [28]

5.4. Opportunity Fuels Related to Herbaceous Biomass

A host of additional biomass fuels have been fired in boilers and other energy systems; many of these fuels are related to the agricultural community and to herbaceous biomass. These materials range from the various manures produced by cattle, swine, poultry, and other animals to products of rendering processes. The manures are becoming of significance in the opportunity fuels arena, as regulatory pressures force consideration of their use in boilers and gasifiers. Table 5.15 documents characteristics of typical manure opportunity fuels.

The manures are among the most difficult opportunity fuels from the biomass community. They tend to be high in both nitrogen and ash, and tend to have highly reactive ash compositions as shown in Table 5.16.

Table 5.15. Proximate and Ultimate Analyses for Typical Manures[1]

Parameter	Fuel				
	Dairy tie-stall manure	Dairy free-stall manure	Poultry Litter	Sheep manure	Swine waste[2]
Moisture %	64.7	69.8	20.0	47.8	97.8
Proximate Analysis (wt %, dry basis)					
Volatiles	76.0	30.1	55.3	65.2	59.6
Fixed Carbon	18.1	7.4	7.7	14.0	7.3
Ash	6.0	62.5	17.0	20.9	33.1
Ultimate Analysis (wt %, dry basis)					
Carbon	48.6	22.6	38.1	40.6	38.0
Hydrogen	5.8	2.9	5.6	5.1	5.5
Nitrogen	1.4	1.1	3.5	2.1	3.2
Sulfur	0.1	0.1	0.6	0.6	0.6
Oxygen	38.1	10.8	34.9	30.7	19.6
Ash	6.0	62.5	17.0	20.9	33.0
HHV (MJ/kg)	19.04	8.46	14.85	16.00	17.00
HHV (Btu/lb)	8203	3644	6399	6895	7328

Notes: [1]Totals may not add to 100.0 due to rounding. [2]Swine waste is liquefied for transportation and disposal. Sources: [24, 34, 35]

The use of manure remains largely restricted to anaerobic digestion systems, although FibreWatt and others have developed stoker fired systems for poultry litter, and increased emphasis is being placed on gasification of these materials.

5.5. Conclusions

The herbaceous opportunity fuels are similar to, yet distinctly different from, the woody biomass fuels. They are commonly used as energy sources in Europe and in China; and they are less commonly used as energy sources in North America. They are lower in bulk density, higher in inorganic matter, and more prone to slagging and fouling that the woody biomass fuels. Further, it is suspected that these biomass fuels

Table 5.16. Chemical Fractionation Analyses of Various Biomass Ash

Ash Constituent	Sheep Manure	Dairy Tie-Stall Manure	Poultry Litter
Potassium			
Water soluble and ion exchange	97	96	41
Acid soluble	1	1	
Insoluble	2	3	20
Sodium			
Water soluble and ion exchange	96	99	90
Acid soluble	2	0	0
Insoluble	2	1	10
Calcium			
Water soluble and ion exchange	58	68	41
Acid soluble	41	31	58
Insoluble	1	1	1
Magnesium			
Water soluble and ion exchange	77	90	69
Acid soluble	21	9	19
Insoluble	2	1	12
Aluminum			
Water soluble and ion exchange	32	12	0
Acid soluble	0	13	0
Insoluble	68	75	100
Silicon			
Water soluble and ion exchange	0	0	0
Acid soluble	0	0	0
Insoluble	100	100	100

Source: [24]

present more significant problems for selective catalytic reduction (SCR) catalysts than the woody biomass fuels.

Demonstrations of firing and cofiring herbaceous biomass fuels have conducted extensively in both Europe and the US. These demonstrations have generally shown the herbaceous biomass energy sources to be significantly more difficult and costly than other opportunity fuels available; however they have demonstrated that herbaceous biomass firing can be accomplished in response to regulatory or incentive-driven markets.

5.6. References

1. Wieck-Hansen, K., P. Overgaard, and O.H. Larsen. 2000. Cofiring coal and straw in a 150 MWe power boiler experiences. Biomass and Bioenergy. 19(6): 395-410.
2. Bransby, D. 2003. Fuel Sources for Cofiring: A Case for Herbaceous Energy Crops in the United States. International Conference on Co-Utilization of Domestic Fuels. Gainsville, FL. Feb. 5-6.
3. Brubaker, G. 2003. Woody and Herbaceous Biomass Production on JEA's Biomass Energy Research Farm. International Conference on Co-Utilization of Domestic Fuels. Gainsville, FL. Feb. 5-6.
4. Mao Jianxiong. 2003. The Energy Structure and the Technology of Co-firing Biomass and Coal in China. International Conference on Co-Utilization of Domestic Fuels. Gainsville, FL. Feb. 5-6.
5. Xie, K.C. 2003. Global Perspectives on Co-Utilization of Fuels. International Conference on Co-Utilization of Domestic Fuels. Gainsville, FL. Feb. 5-6.
6. Miles, T.R., T. R. Miles Jr., L.L. Baxter, R.W. Bryers, B.M. Jenkins, and L.L. Oden. 1995. Alkali Deposits Found in Biomass Power Plants: A Preliminary Investigation of their Extent and Nature. National Renewable Energy Laboratory, Golden, CO.
7. Tillman, D.A. 2002. Cofiring Technology Review. National Energy Technology Laboratory, Pittsburgh, PA.
8. Baxter, L.L. 2003. Biomass Combustion and Cofiring Issues Overview: Alkali Deposits, Flyash, NOx/SCR Impacts. International Conference on Co-Utilization of Domestic Fuels. Gainsville, FL. Feb. 5-6.
9. Tillman, D.A. 2001. Final Report: EPRI-USDOE Cooperative Agreement: Cofiring Biomass with Coal. Contract No. DE-FC22-96PC96252. Electric Power Research Institute, Palo Alto, CA.
10. Amos, W.A. 2002. Summary of Chariton Valley Switchgrass Co-Fire Testing at the Ottumwa Generating Station in Chillicothe, Iowa. National Renewable Energy Laboratory, Golden, CO.
11. Henry, J.F. 1979. *The Silvicultural Energy Farm in Perspective.* in Progress in Biomass Conversion, Vol 1. Academic Press, New York. pp. 215-256.
12. Szego, G. C. and C.C. Kemp. 1973. Energy forests and fuel plantations. CHEMTECH. May.

13. Walsh, M.E., and D. Becker. 1996. Biocost: A Software Program to Estimate the Cost of Producing Bioenergy Crops. Proc. Bioenergy '96. Sep 15-20. Nashville, TN. pp. 480-486.
14. Graham, R.L., W. Liu, H.I. Jager, B.C. English, C.E. Noon, and M.J. Daly. 1996. A Regional-Scale GIS-Based Modeling system for Evaluating the Potential Costs and Supplies of Biomass From Biomass Crops. Proc. Bioenergy '96. Sep 15-20. Nashville, TN. pp. 444-450.
15. Tolbert, V.R., J.D. Joslin, F.C. Thornton, and B.R. Bock. 1999. Biomass Crop Production: Benefits for Soil Quality and Carbon Sequestration. Proc. Fourth Biomass Conference of the Americas. Oakland, CA. Aug 29-Sep 2. pp. 127-132.
16. Costello, R. and H. L. Chum. 1998. Biomass, Bioenergy, and Carbon Management. Proc. Bioenergy '98. Madison, WI. Oct 4-8. pp. 11-18.
17. Wieck-Hansen, K. and P.B. Hansen. 1998. Characterization of Fly Ash from co-combustion of Biomass in a Pulverised Coal-Fired Power Boiler. Proc. 9th Biomass Conference.
18. Rossi, A. 1985. *Fuel Characteristics of Wood and Nonwood Biomass Fuels.* in Progress in Biomass Conversion Vol 5. Academic Press, Orlando, FL. pp. 69-100.
19. Sanderson, M.A., R.P. Egg, and A.E. Wiselogel. 1997. Biomass Losses During Harvest and Storage of Switchgrass. Biomass and Bioenergy. 12(2): 107-114.
20. Arlington, W. 1977. *Bagasse as a Renewable Energy Source.* in Fuels and Energy from Renewable Resources. Academic Press, New York. pp. 249-256.
21. Lalor, W.F. 1977. *Use of Ginning Waste as an Energy Source.* in Fuels and Energy from Renewable Resources. Academic Press, New York. pp. 257-274.
22. Folkedahl, B.C., C.J. Zygarlicke, P.N. Hutton, and D.P. McCollor. 2001. Biomass for Energy—Characterization and Combustion Ash Behavior. Proc. Power Production in the 21st Century: Impacts of Fuel Quality and Operations. The Engineering Foundation. Oct b28 – Nov 2. Snowbird, UT.
23. Cooper, J., M. Braster, and E. Woolsey. 1998. Overview of the Chariton Valley Switchgrass Project. Proc. Bioenergy '98. Madison, WI. Oct 4-8. pp. 1-10.
24. Miller, B.G., S.F. Miller, C. Jawdy, R. Cooper, D. Donovan, and J. Battista. 2000. Feasibility Analysis for Installing a Circulating

Fluidized Bed Boiler for Cofiring Multiple Biofuels and Other Wastes with Coal at Penn State University. Second Quarterly Technical Progress Report . Work Performed Under Grant No. DE-FG26-00NT40809.

25. Tillman, D.A., B.G. Miller, and D.V. Johnson. 2002. Nitrogen Evolution from Biomass Fuels and Selected Coals. Proc. 19th Annual International Pittsburgh Coal Conference. Pittsburgh, PA. Sep 23 – 26.
26. Prinzing, D.E. 1996. EPRI Alternative Fuels Database. Electric Power Research Institute, Palo Alto, CA. Report TR-107602.
27. Jenkins, B.M., L.L. Baxter, T.R. Miles, Jr., and T.R. Miles. 1996. Combustion Properties of Biomass. Proc. Biomass Usage for Utility and Industrial Power. The Engineering Foundation. Snowbird, UT. April 28 – May 3.
28. Baxter, L.L., T.R. Miles, T.R. Miles, Jr., B.M. Jenkins, D.C. Dayton, T.A. Milne, R.W. Bryers, and L.L. Oden. 1996. the Behavior of Inorganic Material in Biomass-Fired Power Boilers—Field and Laboratory Experiences: Volume II of Alkali Deposits Found in Biomass Power Plants. Sandia National Laboratory, USDOE. Report NREL/TP-433-8142.
29. Tillman, D.A., D.V. Johnson, and B.G. Miller. 2003. Analyzing Opportunity Fuels for Firing in Coal-Fired Boilers. Proc. 28th International Technical Conference on Coal Utilization and Fuel Systems. Coal Technology Association. Clearwater, FL. March 10-13, 2003.
30. Green, A. 2003. Flash Pyrolysis Systematics for the Coalification region. International Conference on Co-Utilization of Domestic Fuels. Gainsville, FL. Feb. 5-6.
31. Tillman, D.A. 2003. Fundamental Fuel Characteristics of Woody Biomass: Implications for Cofiring in Coal-Fired Boilers. International Conference on Co-Utilization of Domestic Fuels. Gainsville, FL. Feb. 5-6.
32. Wiselogel, A.E., F.A. Agblevor, D.K. Johnson, S. Deutch, J.A. Fennell, and M.A. Sanderson. 1996. Compositional Changes During Storage of Large Round Switchgrass Bales. Bioresource Technology. 56: 103-109.
33. Miles, T.R., et. al. 1993. Alkali Slagging Problems with Biomass Fuels. Proc. First Biomass Conference of the Americas. Aug 30 – Sep 2. Burlington, VT. pp. 406-421.

34. Miller, S.F., B.G. Miller, and D. Tillman. 2002. The Propensity of Liquid Phases Forming During coal-Opportunity Fuel (Biomass) Cofiring as a Function of Ash Chemistry and Temperature. Proc. 27th International Technical Conference on Coal Utilization and Fuel Systems. Clearwater Florida, March 4 – 7.
35. Miller, S.F. and B. G. Miller. 2002. The Occurrence of Inorganic Elements in Various Biofuels and its Effect on the Formation of Melt Phases During Combustion. Proc. International Joint Power Generation Conference. Phoenix, AZ. June 24 – 27. (IJPGC2002-26177).
36. Unterberger, S., C. Lopez, and K.R.G. Hein. 2001. EU-Project "Prediction of Ash and Deposit Formation for Biomass Pulverized Fuel Co-Combustion. Proc. Power Production in the 21st Century: Impacts of Fuel Quality and Operations. The Engineering Foundation. Oct b28 – Nov 2. Snowbird, UT.
37. Kaer, S.K. 2001. Modeling Deposit Formation in Straw-Fired Grate Boilers. Proc. Power Production in the 21st Century: Impacts of Fuel Quality and Operations. The Engineering Foundation. Oct b28 – Nov 2. Snowbird, UT.
38. Korbee, R., J. Kiel, M. Zevenhoven, B Skirifvars, P. Jensen, and F. Frandsen. 2001. Investigation of Biomass Inorganic Matter by Advanced Fuel Analysis and Conversion Experiments. Proc. Power Production in the 21st Century: Impacts of Fuel Quality and Operations. The Engineering Foundation. Oct b28 – Nov 2. Snowbird, UT.
39. Frandsen, F.J., P.A. Jensen, W. Lin, and K.Dam-Johansen. 2001. Practical Experience with Ash and Deposit Formation in Danish Biomass-Fired Boilers. Proc. Power Production in the 21st Century: Impacts of Fuel Quality and Operations. The Engineering Foundation. Oct b28 – Nov 2. Snowbird, UT.
40. Tillman, D.A. 1994. Trace Metals in Combustion Systems. Academic Press, San Diego, CA.
41. Envirosphere Co. 1984. Final Biomass Ash Study. Prepared for California Energy Commission. Contract Number 500-81-037.
42. Tillman, D.A. 1991. The Combustion of Solid Fuels and Wastes. Academic Press. San Diego, CA.
43. NCASI. 1984. National Council of the Paper Industry for Air and Stream Improvement Technical Bulleting #447. New York.

44. Overgaard, P., B. Sander, and N.O. Knudsen. 1999. Cofiring of Coal and Straw. Proc. Fourth Biomass Conference of the Americas. Oakland, CA. Aug 29 – Sep 2. pp. 1299-1306.
45. Bush, P.V., D.I. Bransby, H.A. Smith, C.R. Taylor, and D.M. Boylan. 2001. Evaluation of Switchgrass as a Co-Firing Fuel in the Southeast. Final Technical Report. Southern Research Institute, Birmingham, AL. For USDOE Cooperative Agreement No. DE-FC36-98GO10349.
46. Zemo, B., D. Boylan, and J Eastis. 2002. Experiences Co-firing Switchgrass at Alabama Power's Plant Gadsden. Proc. 27th International Technical Conference on Coal Utilization and Fuel Systems. Clearwater, FL. March 4-7. pp. 1109-1120.
47. Boylan, D., V. Bush, and D.I. Bransby. 2000. Switchgrass cofiring: pilot scale and field evaluation. Biomass and Bioenergy. 19(6): 411-418.
48. Aerts, D. and K. Ragland. 1997. Final Report: Cofiring Switchgrass in a 50 MW Coal-Fired Boiler. Report to the Great Lakes Regional Biomass Energy Program, the Electric Power Research Institute, Madison Gas and Electric Company, Wisconsin Power and Light Company, and Nebraska Public Power District. Under Contract CGLG-95-008 and EPRI Agreement RP-4134-05.

[illegible] and M.A. Kimbrell. 1998. [illegible] Proc. [illegible] Conference of the [illegible] pp. 1384–1386.

[illegible] Southern [illegible] Birmingham, AL. [illegible] Agreement No. [illegible]

[illegible] 2002. Experiences Co-firing [illegible] Plant Gadsden [illegible]

[illegible]

[illegible] Agreement No. [illegible]

CHAPTER 6: TIRE-DERIVED FUEL AS AN OPPORTUNITY FUEL

6.1. Introduction

Used tires from automobiles, trucks, tractors, and other mobile equipment provide an energy resource of significant interest to many utilities. Tires—and tire-derived fuel (TDF)—have a high calorific value along with other favorable fuel characteristics. At the same time they present material preparation and handling issues for fuel users [1,2]. For environmental reasons, they are increasingly difficult and costly to dispose of in landfills. In 1990, only 25 million tires or 11% of the annually generated scrap tires in the U.S. were utilized (recycled, retreaded, and burned for energy). In 1994, this number increased to 138 million tires or 55% of the annually generated scrap tires with the largest increase due to tires used for energy (101 million tires) [3]. The trend towards increased use of tires in material and energy applications continues its upward path. However, with an estimated one to three billion tires in stockpiles and landfills throughout the United States, this potential energy source is enormous [4 - 6]. This source of energy also is significant in Europe, Japan, Korea, and numerous other industrialized economies where transportation and mechanized agricultural systems are well developed.

This chapter reviews the fuel characteristics of TDF, along with several commercial demonstrations of tire-derived fuel cofired with coal in industrial and utility furnaces. Included in the fuel characteristics section are discussions of fuel characteristics, preparation and handling systems of the tire-derived fuel, methods of utilization of the cofired fuel including appropriate combustion systems (e.g., cyclone boilers, stokers, fluidized bed boilers) and environmental results of the TDF cofiring programs.

It is important to consider the fundamental properties of tires used by automobiles, trucks, tractors, off-road construction equipment, and other mobile machinery. It is interesting to note that the leather-covered pneumatic tire was invented in 1845 by Robert W. Thompson. John B. Dunlop among others improved on Thompson's invention and in 1895 pneumatic tires appeared on automobiles.

Rubber rapidly replaced leather in pneumatic tire manufacture. Originally, natural rubber – polyisoprene – was the basis of tire manufacture. Rubber is defined in ASTM D1566 as:

> ". . . a material that is capable of recovering from large deformations quickly and forcibly, and that can be or already is modified to a state in which it is essentially insoluble." [7].

Crude rubber, $(C_5H_8)_x$, an unsaturated compound is vulcanized with sulfur to achieve cross linkages and a more stable compound. Vulcanization is an irreversible process during which a rubber compound, through a change in its chemical structure, e.g., cross-linking, becomes less plastic and more resistant to swelling by organic liquids. The result is that elastic properties are conferred, improved, or extended over a wide range of temperatures. Vulcanized rubber has a high tensile strength, is insensitive to temperature fluctuations, and can have significant abrasion resistance when compounded with carbon black [8]. Vulcanization processes applied to rubber compounds typically utilize sulfur and zinc.

While natural rubber began as the basis for automobile tires, synthetic rubber products replaced natural rubber – partly as a consequence of the rubber shortage of World War II. Synthetic rubber products include styrene-butadiene rubber (SBR), polybutadiene rubber (BR), polyisoprene rubber (IR), nitrile rubber, neoprenes, polysulfides, polyacrylate rubber, and a host of other products with 65% of all SBR manufactured being used in tire production. [7]. SBR has the following chemical structure [9] and Table 6.1 lists some of the major properties of SBR vulcanized with 50% carbon black [10].

$$-\left[CH_2-CH=CH-CH_2\right]_m-\left[CH(C_6H_5)-CH_2\right]_n- \qquad (6.1)$$

Table 6.1 Selected Properties of Vulcanized SBR

Property	Value
Density	1.5 g/cm^3
Heat Capacity	1.5 kJ/kg-K
Thermal Conductivity	0.300 W/m-K

Source: [10]

Tires are made from combinations of these synthetic rubber compounds, particularly SBR, BR, IR, halogenated butyl rubber, and ethylene-propylene rubber. These synthetic rubbers are used because of the resistance to abrasion, reduced rolling resistance, resistance to cracking and to crack growth, adequate flexibility at low temperatures, a high coefficient of friction between the tread and the road to minimize slipping and skidding, sufficient stability of the material with time so that excessive deterioration does not occur in the normal life of a tire, and an ability to not develop excessive temperatures in service [7]. Table 6.2 lists some of the properties of these rubber compounds used in tire manufacturing.

Significant quantities of carbon black (~97% pure carbon) are used to improve the properties of the tire rubber. Typically this carbon black is made in furnaces firing highly aromatic oil feedstocks free of coke, and low in asphaltene content [11]. Carbon black reinforcement of rubber was discovered in 1912, and has traditionally been used to impart strength, shape, load carrying capabilities, and bruise and fatigue resistance to the tires. Tire cords originally were cotton. Subsequently rayon, nylon, polyester, glass and steel became cord materials of choice. In the United States, steel is the dominant cord material [12].

Table 6.2. Comparisons of Types of Rubber Used in Tire Production[1]

	NR	SBR	IR	BR
Physical Properties				
Flame Resistance	X	X	X	X
Heat Resistance	C	C	C	C
Cold Resistance	B	B	B	B
Mechanical Properties				
Tensile Strength @ 7 MPa	>21	>14	>21	>14
Hardness, Shore A	30	40	30	40
Cold Rebound	A	B	A	A
Hot Rebound	A	B	A	A
Tear Resistance	A	C	A	C
Abrasion Resistance	A	A	A	A
Chemical Stability				
Sunlight Aging	C	C	C	C
Oxidation Resistance	B	B	B	B
Natural Gas	X	X	X	X
Dilute Acids	B	B	B	B
Concentrated Acids	C	C	C	C
Specific Gravity	0.92	0.93	0.92	0.94

Notes: [1]NR=natural rubber, SBR=styrene butadiene rubber, IR=polyisoprene rubber, BR=butadiene rubber. Ratings: A=excellent, B=good, C=fair, X=poor. Source: Data from [7]

Modern tires, then, are combinations of synthetic rubber compounds, carbon black and cord material brought together into a high strength product capable of significant load support and significant resistance to stress. Because of the materials used, they are highly reactive, exhibit significant aromaticity as a consequence of the carbon black, and have substantial concentrations of such metals as zinc, resulting from the vulcanization process. The cords or belting material may or may not increase the inorganic matter in the tires, depending upon the cord material of choice. When such products wear out, as a consequence of use, they are available for conversion into fuel, crumb rubber, playground tire chips, paving material or other uses.

6.1.1. Overview of the TDF Production Processes

This section describes the preparation required to produce a TDF fuel product that is compatible with large-scale energy production facilities. In essence this section discusses machinery required to totally destroy the integrity of a high strength product that Messrs. Thompson, Dunlop, Goodyear, Firestone, and Goodrich dedicated their entire lives to develop!

6.1.1.1. General Description of Tire-Derived Fuel

Tire-derived fuel is broadly classified as a fuel feedstock derived from automobile, truck, off-road and specialty tires. These tires are either used whole or chipped into sized pieces for blending with other more conventional fuels such as coal. They may contain a significant amount of steel used in reinforcement of the sidewall and rim. Depending upon production requirements, this steel may need to be removed from the tire.

The great advantage of TDF as an opportunity fuel is its high calorific value, low ash (if the steel is removed) and low to moderate sulfur content. This makes its usage in boilers attractive if it can be prepared and fed into the boiler in an economic manner.

There are three basic types of tire-derived fuel in use today: TDF with steel, without steel and crumb rubber. Tires come from predominantly passenger automobiles and trucks; however, on a per mass basis tractor, sport utility vehicle, and commercial truck tires provide a significant amount of TDF [13].

TDF with Steel.

Tire-derived fuel with steel implies utilizing the entire tire as is or shredded into sized pieces. There is essentially no processing done on these tires except that the steel rim may be removed. This fuel has the highest ash content and the lowest calorific value of the TDF products. This is the most economical from a processing perspective but can result in utilization problems with the additional iron present in the ash. Other utilization problems occur with materials handling. Steel wire, randomly liberated during the shredding process or in materials handling, can cause damage to coal handling belts, primary and secondary crushers, and to

other elements of the feed system. Further, steel "whiskers" on the tire chips can cause significant bridging in material handling systems – bridging that is difficult to break up due to the interweaving of fuel particles.
The coarse shredded TDF is typically produced in the largest particle sizes. Common sizes are <38 – 50 mm (<1.5 – 2 in.). Periodically it is shredded to smaller particle sizes as well.

TDF without Steel.

In order to make the TDF fuel more compatible with energy production facilities, it is desirable to remove the steel rim and bead material. This is usually accomplished by removing the rim and steel bead from the tire sidewall and magnetically removing the steel. While most of the steel is removed, some still remains within the tire chip and may present handling issues. Manufacturing wireless chips requires additional machinery. However, such chips are periodically produced at particle sizes of <15 – 25 mm (0.6 – 1 in.).

Crumb Rubber.

Crumb rubber is TDF essentially without any steel. This is usually material chipped/ground to less than 6 mm and passed through magnetic separators to remove any remaining steel. This material has the highest calorific value, lowest ash content but is the most expensive to process. It is mainly used in specialty applications but not for energy production.

6.1.1.2. Tire-Derived Fuel Production Statistics

Scrap tires comprise nearly one percent of solid municipal waste in the United States and yet could provide a significant resource to many utilities. Unfortunately, landfilling has been and continues to be the most common tire disposal option. However, like many other wastes, the increase in landfill environmental regulations and increases in tipping fees have led the industry to using other options. The estimated number of tires in stockpiles within the United States ranges between 1 and 3 billion tires and approximately one scrap tire per person is generated every year

[14]. In 1998 there were approximately 2,800 individual stockpiles located throughout the U.S. [6].

In 1990, only 25 million tires or 11% of the annually generated scrap tires in the U.S. were utilized (recycled, retreaded, burned for energy). In 1994, this number increased to 138 million tires or 55% of the annually generated scrap tires [15]. Figure 6.1 shows the distribution of scrap tire usage in the United States in the mid-1990s, while Figure 6.2 shows the distribution of TDF in the fuel market between 1998 and projected for 2003 [15].

In Japan, where energy costs are high, more than 50% of the 102 million scrap tires generated annually are prepared as TDF; however this is simply incineration [6]. Incineration is expected to come under scrutiny as Japan is pressured into reducing emissions, especially greenhouse gases. The distribution of scrap tire usage in Japan in shown in Figure 6.3. In Europe, scrap tire uses vary. In Britain 65% of the scrap tire production, or 15 million tires annually, are landfilled or abandoned with scrapped vehicles. Only about 5 million tires annually are utilized as TDF [6]. The distribution of tire usage for Britain is illustrated in Figure 6.4.

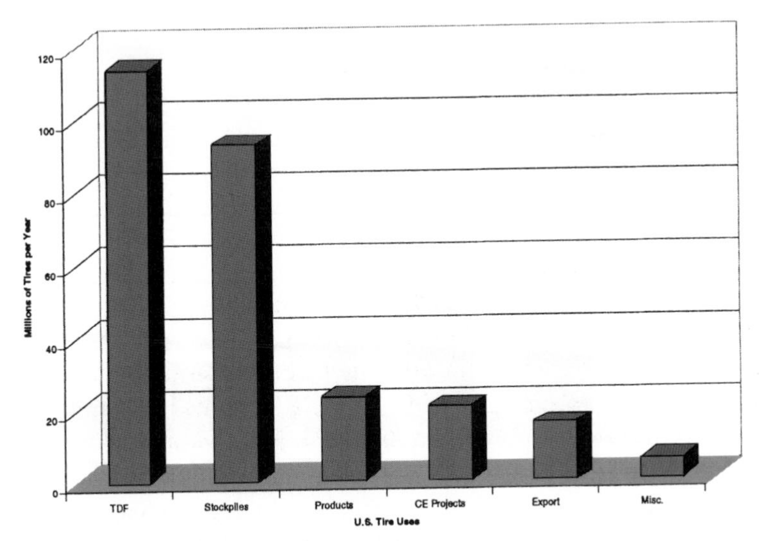

Figure 6.1 United States scrap tire usage
(Source: Data from [15]).

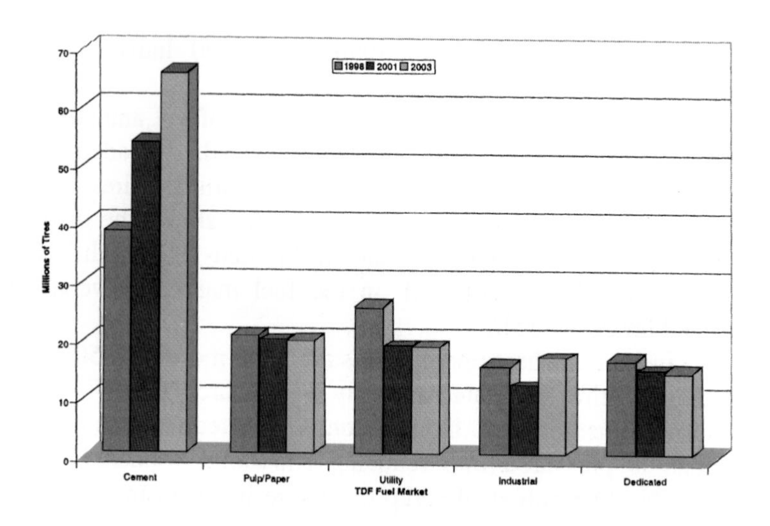

Figure 6.2 TDF usage trends in the US market
(Source: Data from [15]).

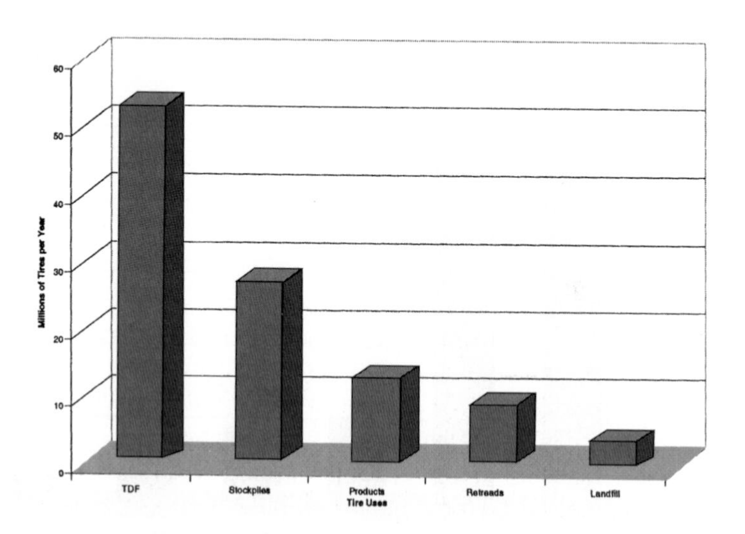

Figure 6.3. Japan scrap tire usage
(Source: Data from [6]).

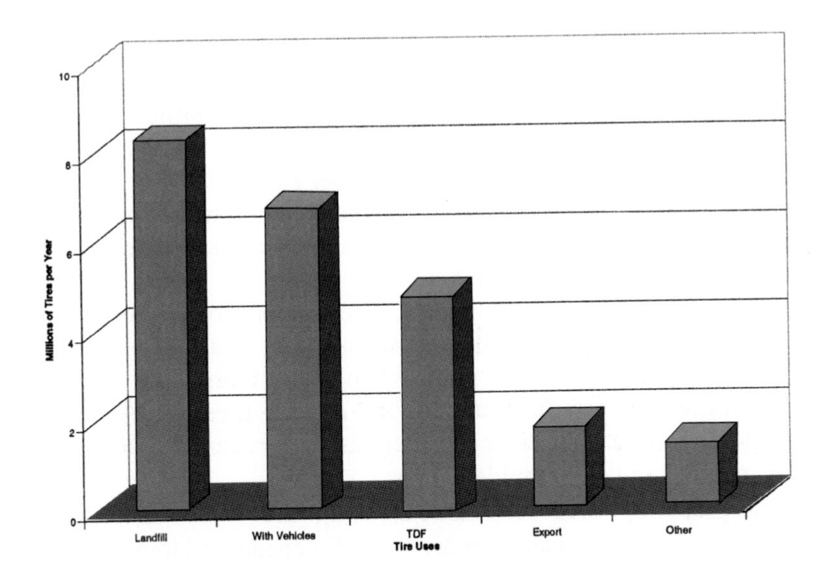

Figure 6.4. Scrap Tire Usage in Great Britain
Data from [6]

6.1.1.3. Uses of Tire-Derived Materials

Current options for waste tires are expanding. Some common uses are tire-derived fuel, civil engineering applications, manufactured products and pyrolysis [13]. Tire-derived fuel (TDF) is usually produced by cutting waste tires into small chips and removing the wire from them. Whole tires are used for playground equipment, reef construction, soil erosion, retreads, crash barriers, and for a fuel in specifically designed tire-fired boilers. General construction practices use tires for erosion control, retaining walls, as layered thermal insulation that can help reduce frost around basements, as a drainage aid on highway edges, and leachate liners for septic tanks and landfills, and blasting mats.

Waste tires, whole or chipped, are used in cement kilns, lime kilns, paper mills, utility boilers, industrial boilers, iron foundries, copper smelters, and tire-derived fuel oil to supplement coal. TDF typically refers to shredded tires in the form of rubber chips that are burned with coal in utility coal-fired boilers, or with coal in cement kilns. Scrap tires have been used as a supplemental fuel source in Japan, Europe, and the Untied States since the 1970s. According to the Rubber Manufacturers Association of Washington, D.C., approximately 125 million tires are used annually for tire-derived fuel [15]. TDF compares favorably to the burning of coal in every aspect except for higher zinc emissions.

Waste tires and associated products are used in highway construction, as rubber-modified asphalt, chip and seal coatings, and sub-base material for roads over soft or marshy areas.

Miscellaneous uses include chopped, shredded or ground tires used in smaller rubber items such as floor mats, sandals and molded rubber objects. Cut and punched tires are used to make items such as belts, gaskets, muffler hangers and agricultural-related tires. Newer uses include polymer-treated rubber powder used for carpet backing or underlay, roofing and other molded items.

In preparing tires for markets, scrap tires may be chopped, shredded or ground. Shredded tires are typically cheaper to transport than whole tires. Costs to get tires to a market include the cost of transportation, labor, and shredding which is approximately $0.50-0.90 per tire. This cost is offset by tipping fees which are typically in the $70-$80/ton range. The markets for scrap tires are numerous in the U.S. but dominated by the cement industry with a 27% market share. The Scrap Tire Management Council estimated that the products market consumed roughly 15 million tires for 1996 [15].

The Chicago Board of Trade may begin listing some of these waste tire material commodities that have been publicly offered for trade. A barrier to listing waste tire material as a commodity has been the lack of industry standards. Specifically, questions regarding the quality of scrap tire crumb rubber have limited this commodity's acceptance in the marketplace. To counter this problem, the American Society of Testing and Materials has developed a set of scrap tire crumb rubber specifications. On August 10, 2001 the ASTM Committee D34 on Waste Management issued a new standard for material recovery of scrap tires for fuel, D6700, Standard Practice for Use of Scrap Tire-Derived Fuel [16,17]. This standard provides definitive standards for size, distribution, sampling, and testing of tire-derived fuel. According to the document, it "covers and provides guidance for the material recovery of scrap tires for their fuel value. The conversion of a whole scrap tire into chipped form for use as a fuel produces a product called tire-derived fuel (TDF). This recovery practice has moved from a pioneering concept in the early 1980s to a proven and continuous use in the United States with industrial and utility applications." It should be noted that whole tire combustion for energy recovery is not discussed in this standard since whole tire usage does not require tire processing to a defined fuel specification.

6.1.2. Tire-Derived Fuel Production

There are basically two types of tire shredding equipment in commercial use, the solid cutter shredder and the replaceable cutter tire shredder [18,19]. Both of these machines shear a whole tire to produce a chip with a fixed size in one dimension, but to control the other chip dimension, additional passes through the chipper may be required.

6.1.2.1 Solid Cutter Shredder

The solid cutter shredder is a low-speed shear shredder that lends itself to many shredding applications including: tires, municipal solid waste, oversized bulky waste, demolition and construction debris, etc. Figure 6.5 shows the principle behind the solid cutter shredder. Two counter-rotating shafts with intermeshing cutters are used to grab the whole tire and force it through the pinch points between opposing cutters. The particle size of the tire chips is controlled by basically two factors – cutter width and number of hooks on each cutter. A cutter with a 25 mm cutter disc will produce a 25 mm x ? mm chip since the length dimension will be determined by the distance between successive hooks on the cutter perimeter. To get a controlled chip size, screening and multiple passes through the shredder may be required.

Advantages of the solid cutter shredder include flexibility – the solid cutter shredder is a multi-purpose unit that can shred many types of materials; simple design – the cutting box arrangement basically consists of two shafts and cutters that can be easily maintained in the field. Also, the cutters can be removed and resharpened to a distinct edge several times; variable chip size – because the cutters are slid onto each shaft, they can easily be changed to produce different size chips; and bulk feeding capability – many different feedstocks can be fed in bulk be grapple, front end loaders, conveyors, etc. directly into the shredder hopper.

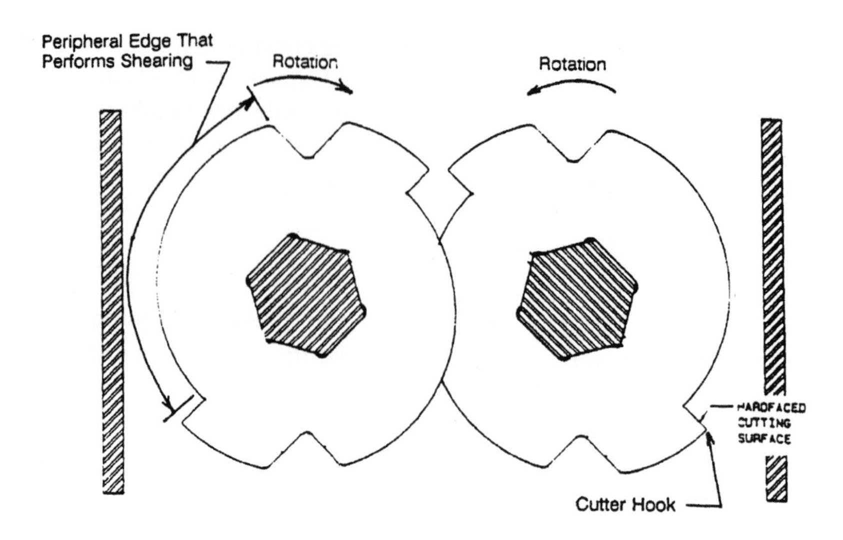

Figure 6.5. Schematic of solid cutter shredder shearing blades
(Source[19] with permission).

The two main disadvantages of the solid cutter shredder are that it was not specifically designed for tires and thus tends to tear the tire rather than shear or cut it, and secondly the sharpness of the cutter edge tends to deteriorate rapidly. This can be overcome with the use of harder steels; however they are more costly.

6.1.2.2 Replaceable Cutter Tire Shredder

The second type of tire chipper is the replaceable cutter tire shredder. This is a low-speed shear shredder specifically designed for tires. The principle cutting feature is shown in Figure 6.6. Note the addition of star-shaped feed rolls that force the individual tires, one at a time, into the cutting zone. As the tires are fed into the cutting zone, the cutters engage them with a true shearing action.

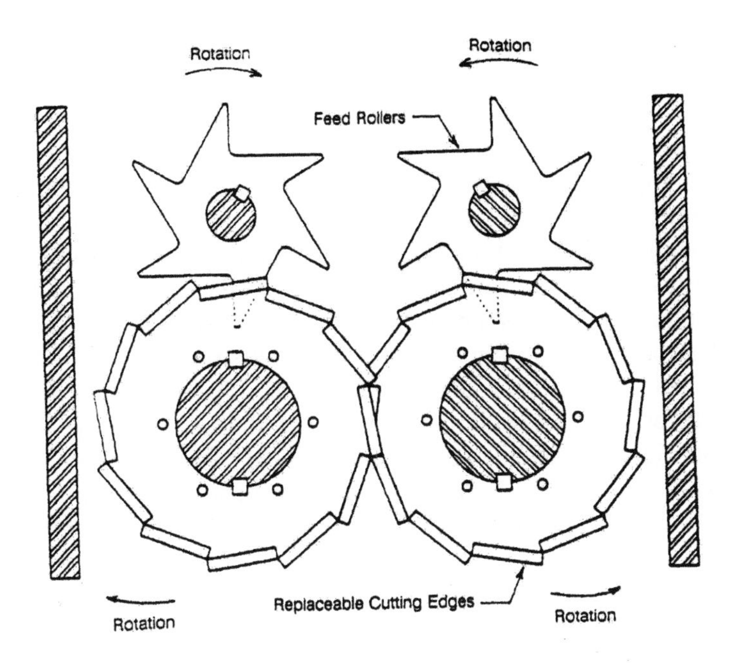

Figure 6.6. Schematic of replaceable cutter tire shredder system
Source: [19] with permission.

As with the solid cutter shredder, one dimension of the tire chip is controlled by the width of the cutter segments. However, because there are not hooks, the other dimension may vary up to about 0.25 – 0.30 m in length. As a result, multiple passes are required to obtain a uniform chip size. Advantages of the replaceable cutter tire shredder include the following: this type of shredder is specifically designed for tire shredding and thus produces a higher quality cut and more uniform size chips; and the removable cutter blades are made from a higher quality steel alloy and can be replaced without the need to replace an entire cutter assembly. Some of the disadvantages of the replaceable cutter tire shredder are that it requires only one tire at a time into the feed rollers and the chip size cannot be varied easily. Knife maintenance, however, is a critical issue for production of TDF in all systems.

6.1.2.3. Other Types of Tire Derived Fuel Production Equipment

This equipment includes alternatives to shredding, along with wire removal systems. Recently alternative shredding/particle size reduction equipment has been developed. Of particular interest is the development of the hydroblasting system where the rubber is removed from the belting and bead wire with high pressure water. The water not only separates the rubber from the beat wire and other cord material, but it also produces tire chips at a particle size of 13 – 25 mm (1/2 – 1 in.).

The bead wire in tires consists of many strands of high tensile strength steel to reinforce the tire sidewalls. Special equipment is available to cut the sidewalls off the tire and thus increase the life of cutter blades in the shredding equipment. However, the current procedure for removing the tire bead is very labor intensive and expensive. In most applications it is more economical to sharpen/replace cutter blades than to debead a tire.

Passenger car tire rims are typically removed with the "three-pronged" hydraulic device that simultaneously crushes the rim allowing the tire to be lifted off. A similar, but more involved procedure is used with larger off-road and truck tires. These procedures are also very labor intensive and costs must be balanced against sharpening/replacing cutter blades.

6.1.2.4 Cost Information for Tire Chip Preparation

Processing costs for a tire shredding operation will vary depending upon many factors, such as feed material, throughput, system being used (portable vs stationary); site conditions, preventative maintenance program, disposal options, power available; and local markets, desired end product, geographic location.

Perhaps the key to a successful tire chipping operation is properly setting the processing tipping fees, which must include tire processing, disposal and profit. Fees must be evaluated on a "per tire" versus "per ton" basis; this will depend on the amount of processing to be done on the tires and the desired end product. In the mid-1990s, typical tipping fees were between $70-$80 per ton except in those areas where land disposal costs were extremely high. Table 6.3 provides some estimates of the capital cost of different tire shredding equipment based on throughput and system configuration.

Table 6.3 Capital costs of various tire shredding systems

Shredder Type/ Manufacturer	Estimated Costs ($1,000)	System Configuration	Estimated Throughput (kg/s)
		Replaceable Cutters	
Columbus McKinnon	$500 - $525 $435 - $460	Portable Stationary	3.0-3.3 - Rough Shred 2.0-2.5 – 50mm 1.0-1.3 – 25mm
Triple S/Dynamic	$475 - $500 $400 - $425	Portable Stationary	3.0-3.3 - Rough Shred 2.0-2.5 – 50mm 1.0-1.3 – 25mm
		Rotary Shear	
Eidal	$400 - $425 $290 - $315	Portable Stationary	2.5-3.0 - Rough Shred 1.5-2.0 – 50mm 0.8-1.0 – 25mm
ERS	$500 - $525 $425 - $450	Portable Stationary	3.0-3.3 - Rough Shred 2.0-2.5 – 50mm 1.0-1.3 – 25mm
Mac-Saturn	$400 - $425 $340 - $365	Portable Stationary	2.5-3.0 - Rough Shred 1.5-2.0 – 50mm 0.8-1.0 – 25mm
Mitts & Merrill (Carthage)	$400 - $425 $250 - $275	Portable Stationary	2.0-2.5 - Rough Shred 1.3-1.8 – 50mm 2.5-3 – 25mm
Shredding Systems	$450 - $475 $375 - $400	Portable Stationary	2.5-3.0 - Rough Shred 1.5-2.0 – 50mm 0.8-1.0 – 25mm

Source: [19]

6.1.3. Advantages of Tire-Derived Fuel

The use of tire-derived fuel as a cofiring feedstock in power generation has many advantages. First, because of the high calorific value of TDF, particularly the wire-free TDF, the cost of TDF in ¢/GJ (or $\$/10^6$ Btu) is lower than that of any fossil fuel – with the possible exception of Powder River Basin coal delivered locally or to a mine-mouth plant – and is competitive with even the lowest cost biomass fuels. This, combined with the lower ash content of TDF (without steel) makes it a viable blending fuel. In addition, the sulfur content of TDF is less than most eastern bituminous coals and comparable to medium-sulfur coals throughout the world.

By comparing the TDF fuel with nearly all other fossil fuels on a kg/GJ basis, the TDF is an attractive blending fuel for most boiler applications. The availability of a tire source to produce TDF within a local area and transportation costs to deliver the fuel to the boiler site need to be considered when specifying its use.

6.1.4. Reality of Using Tire-Derived Fuel

The use of TDF is, unfortunately, not without concerns. In most instances, the steel wires within the tire chip cause problems with large-scale fuel handling equipment. These wires intertwine and form large agglomerates that plug handling equipment and even tear conveyor belts. The steel wire within a tire chip may also be a concern in the combustion of the TDF as the additional iron may result in increases in corrosion rates and slagging on boiler waterwalls. Other constituents in the ash such as zinc or chromium may also be detrimental to overall boiler efficiency.

Finally, the overall cost of producing an acceptable product must be considered. While most tire chippers can produce a uniform sized product in one dimension, the other dimension can vary significantly. Therefore, it becomes an economic decision as to whether multiple passes will be required through a chipper to produce a consistent, two-dimensional product. Screening, to verify size consist, may have to be done at the end-user site to reduce problems associated with burning TDF.

6.1.5. Basis of Consideration

The remaining sections of this chapter focus on the fuel characteristics of TDF including some comparisons to other types of fossil fuels. The bulk of the chapter focuses on case studies where TDF has been utilized in industrial and utility boilers. These studies are highlighted to provide information on successes and concerns experienced by the various users.

6.2 Fuel Characteristics of Tire-Derived Fuel

Tire-derived fuel used as a fuel supplement comes in three main forms. In some cases, whole tires have been injected into a boiler and burned. Others require that tire crumb be produced without any steel and less than about 6 mm size. The most common size for utilizing TDF is between 25 mm and 50 mm squares with the steel removed. The economics of chipping to finer sizes and removing the steel are very important in determining the long-term potential for TDF usage. This section provides typical analyses of the TDF, including those analyses required for fuel specification in combustion systems.

6.2.1. Proximate and Ultimate Analysis of Tire-Derived Fuel

Table 6.4 lists typical proximate, ultimate and calorific value analyses of TDF samples [20]. As expected, those samples with steel had a lower heating value and higher ash content than those without the steel. However, the calorific value, even for those samples with steel, is higher, in most cases, than coal. This is one of the attractive features of the TDF fuel; because of its higher calorific content it can replace coal with actually less mass being fed into the boiler. Even more pronounced is the low ash content of the samples without steel. The sulfur content is typical of eastern bituminous coals, while the nitrogen content is considerably lower. However, on a kg/GJ basis both the sulfur and nitrogen are lower than most coals – potentially resulting in lower emissions.

Using these typical analyses, several fundamental derived parameters can be calculated. These parameters are listed in Table 6.5 and are used to estimate combustion propensity and reactivity. The values

Table 6.4 Typical Proximate and Ultimate Analyses of TDF

Parameter	TDF with Steel	TDF w/o Steel
Proximate Analysis (wt %)		
Moisture	0.75	1.15
Volatile Matter	54.23	63.50
Fixed Carbon	21.83	29.28
Ash	23.19	6.22
Ultimate Analysis (wt %)		
Carbon	67.0	81.70
Hydrogen	5.81	7.18
Nitrogen	0.25	0.56
Sulfur	1.33	1.62
Oxygen	1.64	2.65
Ash	23.19	6.29
Higher Heating Value		
MJ/kg	31.06	36.68
Btu/lb	13,362	15,781

Source: Data from [20]

determined are the volatile-to-fixed carbon ratio, hydrogen/carbon ratio, oxygen/carbon ratio, and the kg sulfur/GJ heat input values.

The reactivity shown in Table 6.5, reflected particularly in the hydrogen/carbon atomic ratios and the VM/FC ratios, is consistent with the structure shown in Figure 6.1. Note that the VM/FC ratio exceeds 2.0 for both TDF with steel and TDF without steel, and the H/C atomic ratios exceed unity. This is consistent with the structure of SBR, which is low

Table 6.5. Fundamental Derived Values for TDF Samples

Parameter	TDF w/ Steel	TDF w/o Steel
Reactivity Measures		
V/FC	2.48	2.17
H/C Atomic	1.083	1.07
O/C Atomic	0.018	0.024
Sulfur Concentrations		
kg S/GJ	0.43	0.44
kg SO_2/GJ	0.86	0.88

Source: From Table 6.4.

in aromaticity. The structure of SBR has a calculated aromaticity of 50 percent—well below that of even the highly volatile subbituminous coals. Recognizing that TDF contains significant concentrations of carbon black, and that the carbon black is highly aromatic, the actual aromaticity of TDF is probably around 65 percent. This is consistent with the aromaticity of many western US subbituminous coals and North Dakota lignites.

6.2.2 Inorganic (Ash) Constituents

Table 6.6 lists some of the major ash constituents in various TDF samples. Of particular note is the very high zinc content of the ash resulting from the vulcanization processes. As mentioned earlier, the emissions of TDF compared to coal can be either the same or lower with the exception of zinc emissions. It is also important to note that the iron content in the ash is high, even for the wireless tires. This iron has the potential to flux slag formation in cyclone boilers, particularly if the iron is finely divided (e.g., not in bead wire). The high iron content is also beneficial when TDF is used in cement kilns, since it reduces the amount of iron additive needed to obtain cement with the proper iron oxide content.

Table 6.6 Ash constituents for TDF samples

Parameter (wt % of ash)	TDF w/ Steel	TDF w/o Steel
Al_2O_3	1.12	7.85
SiO_2	4.13	20.95
TiO_2	0.26	7.40
Na_2O	0.16	0.81
K_2O	0.20	0.72
Fe_2O_3	80.10	23.05
MgO	0.38	0.99
CaO	1.80	3.61
ZnO	8.96	25.70
Other	0.21	1.62

Source: Data from [12]

6.2.3 *Trace Element Emissions*

The EPA Office of Air Quality Planning and Standards completed a study on trace element emissions from TDF burning facilities [21, 22]. The data were from 22 industrial facilities that have used TDF: 3 kilns (2 cement and 1 lime) and 19 boilers (utility, pulp and paper, and general industrial applications). In general, the results indicate that properly designed existing solid fuel combustors can supplement their normal fuels (coal, wood, and combinations of coal, wood, oil, petroleum coke, and sludge) with 10 to 20% TDF and still satisfy environmental compliance emission limits. In fact, dedicated tires-to-energy facilities indicate that it is possible to have emissions much lower than produced by existing solid-fuel-fired boilers (on a heat input basis), when properly designed.

In a laboratory test program on controlled burning of TDF, with the exception of zinc emissions, potential emissions from TDF are not expected to be very much different that from other conventional fossil fuels, as long as combustion occurs in a well-designed, well-operated, and well-maintained combustion device. Selected results from this study are shown in Table 6.7. Note that mercury in TDF is typically below detection limits.

Table 6.7 Trace Element Emissions from TDF Facility

Element	Emissions	
	g/MJ	lb/10^6 Btu
Lead	$5.5x10^{-7}$	$1.3x10^{-6}$
Cadmium	$1.6x10^{-6}$	$3.7x10^{-6}$
Chromium	$2.0x10^{-6}$	$4.7x10^{-6}$
Mercury	$2.9x10^{-7}$	$6.7x10^{-7}$
Copper	$3.2x10^{-6}$	$7.5x10^{-6}$
Manganese	$6.9x10^{-7}$	$1.6x10^{-6}$
Nickel	$2.7x10^{-6}$	$6.3x10^{-6}$
Tin	$1.8x10^{-6}$	$4.2x10^{-6}$
Beryllium	$3.1x10^{-5}$	$7.3x10^{-5}$
Zinc	$6.0x10^{-4}$	$1.4x10^{-3}$

Source: Data from [21]

6.3. Cofiring Tire-Derived Fuel in Cyclone Boilers

This section describes several test burns and applications of TDF in cyclone-fired boilers. Cyclone boilers are well suited for TDF applications because they easily handle chipped material and the ash is mainly taken out in fluid form. Thus the additional steel is less of a problem in cyclones than in conventional pulverized-coal boilers.

6.3.1. Cofiring TDF at Allen Generating Station, TVA

Several TDF test burns have been completed at the TVA Allen Fossil Plant outside Memphis, TN [23, 24]. The Allen Station consists of three cyclone boilers, each with the capacity to generate about 272 MW_e of electricity. Early tests cofired up to 8% TDF with coal without significant problems. In addition, in the mid-1990s a series of eleven tests were conducted to evaluate the impact of cofiring and trifiring coal, biomass and TDF. Of these eleven tests, three were baseline (no biomass or TDF), two were 95% coal/5% TDF (mass basis), and two were trifired fuels (85%/10%/5% coal/biomass/TDF and 80%/15%/5% coal/biomass/TDF). These tests investigated the impact of cofiring and trifiring on the fuel yard and fuel handling system, the ability of the plant to achieve full capacity, the impact of alternate fuels on boiler efficiency and consequent net station heat rate, the impact on operating stability and operating temperatures and the effect on emissions.

Table 6.8 provides a summary of the pertinent test results. As can be seen the cofiring of TDF with coal at the 5% level reduced NO_x by about 8%, reduced opacity and resulted in a slight increase in SO_2 emissions. Boiler efficiency was equal to or slightly greater during the cofiring tests.

The results of these tests include the following conclusions. Cofiring and trifiring did not result in any loss of boiler capacity or unit capacity. Cofiring and trifiring did not reduce boiler stability or operability, and did not increase the excess air required for stable and effective operation. Cofiring and trifiring had impacts on boiler efficiency of less than one percent when the unit was firing at full-load and not limited by opacity. Cofiring and trifiring had minimal impacts on

Table 6.8. Summary of Allen TDF Cofiring Tests

Test No.	Coal (%)	Wood (%)	TDF (%)	Load MW	η (%)	NO_x kg/J	SO_2 kg/J	Opacity (%)
2	100	0	0	252	87.9	0.52	0.34	16.3
6	95	0	5	143	88.9	0.47	0.39	3.7
7	95	0	5	270	87.9	0.47	0.34	10.7
8	85	10	5	271	87.7	0.52	0.30	10.1
9	80	15	5	267	87.7	0.52	0.30	9.4
11	100	0	0	272	88.5	0.65	0.30	12.1

Source: Data from [23]

estimated flame temperatures in the barrel. Cofiring and trifiring reduced NO_x emissions relative to all western coal firing; 100% western coal (Test 11) had a NO_x emission of 0.65 kg/J (1.5 lb/10^6 Btu) when firing at MCR and firing under conditions not restricted by opacity; and firing with a blend of 85% coal/10% biomass/5% TDF resulted in a NOx emission of 0.52 kg/J (1.2 lb/10^6 Btu).

Cofiring TDF with western coal did result in an increase in SO_2 emissions; however, a blend of 85% coal/10% biomass/5% TDF achieved about the same level of SO_2 emissions as the coal alone. Opacity certainly did not increase and may have actually decreased during the cofiring and trifiring tests relative to baseline coal-only testing. Ash characteristics were essentially unchanged between the baseline, cofired and trifired tests.

6.3.2. *Cofiring TDF at Baldwin Generating Station, Illinois Power*

Illinois Power conducted an extensive program to test cofiring of TDF at their Baldwin Power Station having two cyclone-fired boilers [25, 26, 27]. Over 11,730 tonnes of TDF were burned during the course of the test program. This represented the ultimate disposal of over 1,293,000 tires. The objectives of these tests were to determine if TDF could be delivered to the boiler reliably, if the TDF could be burned reliably, if TDF posed hazards to the equipment, and if TDF cofiring affected the environmental operation of the plant. Cofiring levels were limited to less than 5% on a heat input basis.

Because the cost to produce 25 mm tire chips is more expensive than 50 mm chips, the initial tests were to see if 50 mm chips could be successfully fed with the coal through the hammer mills. With only about 0.5 tonnes of chips fed, the screens on the hammer mill plugged. Later testing by-passed the hammer mill but still resulted in pluggage of the discharge chutes on the tripper. As a result, the majority of the testing was done with 25 mm chips.

Emission testing was done during the test program; however, because of the small amount of cofiring, no appreciable differences were noted between the baseline and cofiring tests. Leaching tests were also completed on fly ash and slag samples with no apparent differences between the baseline and cofiring tests. The only problem was the presence of unburned pieces of TDF and some wire in the slag. The recommended procedure was to purchase debeaded tire chips.

Carbon measurements made on the baseline ash and the ash from the cofiring tests were similar: 5.52% carbon-in-ash (baseline) and 5.50% carbon-in-ash (cofiring tests). The ESP efficiency varied between 95.6-96.2% during the baseline tests and 93.8-95.7% during the coal and TDF tests. Table 6.9 compares the mineral analyses of the fly ash and slag during the baseline and cofiring tests.

Table 6.9 Ash and Slag Analyses (wt. %)

	Fly Ash		Slag	
Compound	**Coal**	**Coal and TDF**	**Coal**	**Coal and TDF**
SiO_2	50.86	50.15	52.51	53.11
Al_2O_3	20.00	18.71	19.66	19.47
TiO_2	1.31	0.84	0.78	0.70
Fe_2O_3	18.15	18.30	14.73	14.58
CaO	2.99	3.26	6.12	6.70
MgO	1.03	1.11	1.05	1.16
K_2O	2.76	2.62	1.80	1.80
Na_2O	1.28	1.27	0.68	0.72
P_2O_5	0.32	0.02	0.07	0.11
Undetermined	0.88	1.93	1.61	0.09
ZnO	--	0.60	0.10	1.14
Zn (ppm)	2000	483	--	724

Source: Data from [27]

The test program demonstrated that Baldwin Station could successfully cofire TDF as a supplementary fuel at a 2% level capacity. Budget estimates were also prepared for permanent cofiring. These showed an economic benefit of about $200,000/year to the consumers.

6.3.3. Rock River, Wisconsin Power & Light

Wisconsin Power and Light developed a program to cofire TDF with Indiana coal and with PRB coal at the Rock River Generating Station located in Beloit, Wisconsin [28, 29]. This station has two B&W cyclone-fired boilers rated at about 80 MW. The TDF was blended with coal on-site in two ways. WP&L installed its own TDF production facility at the Rock River Generating Station to support this program.

The first approach blended TDF with coal prior to the coal conditioners and the second approach involved bypassing the coal conditioners and adding the TDF onto the main conveyor leading directly to the bunker. Both crumb rubber (nominal 3 mm) as well as nominal 25 mm pieces were able to pass through the conditioners without causing problems.

The initial test was using 36.3 tonnes of crumb rubber and resulted in no major plant impacts. No special changes were made in the coal feed system nor in the combustion cycle for this test. However, the opacity rose from 10-15% to 20-25% and the magnetic pulley on the conveyor showed there was metal in the supposedly wire-free rubber. Subsequent tests were performed with the 25 mm nominal TDF at a 2-10% (by weight) TDF cofiring level. TDF with and without wire was used.

The material and coal handling systems had very few problems throughout the tests. The only major problem was in the gravimetric feeders caused by off-design pieces (very large or rope-like pieces). Unfortunately, when a 25 mm x 25 mm size is specified, it usually means a 12 mm to 50 mm material delivered by the TDF supplier. This led WP&L to construct and operate its own TDF shredding facility to maintain quality control.

The cyclone and boiler operations were not affected by the use of TDF. The TDF caused no new furnace slag accumulations to form; however, when the TDF was larger than 25 mm x 25 mm size, there was not sufficient residence time in the cyclone barrels to completely burn the chips. If one burned TDF larger than 25 mm x 25 mm a maximum of 5% blend worked best. There appeared to be an inverse linear relationship between the TDF chip size and the maximum blend ratio for good unit performance.

The precipitator was affected by the TDF cofiring. It appeared that for each 1% TDF cofired, there was a 1.5% increase in opacity when using the Indiana medium-sulfur coal. However, there were smaller ESP impacts when the TDF was blended with PRB coals.

The fly ash and slag handling systems were not affected by the TDF cofiring. However, the slag produced by the plant did have a significant amount of "bead" wire in it. This caused problems for the slag customers; therefore, WP&L was forced to produce a sized and dewired product using a magnetic separator.

Emissions were extensively tested during the TDF cofiring tests. Table 6.10 lists the results of several emissions tests completed during the program. Note that in these tests the NO_x emissions increased when firing the TDF while the SO_2 emissions were reduced by 27%. Acid emissions were decreased significantly also. Very little changes were noted in trace metal extraction tests on the slag from both the 100% coal and the 95% coal/5% TDF tests.

Based on the results of the TDF tests at Rock River and other WP&L (Alliant) stations, the Edgewater Generating Station of Alliant Energy fires TDF at low percentages to avoid precipitator problems [30].

Table 6.10 Air Emission Results at Rock River Generating Station

Pollutant	Coal	Coal + TDF
SO_2, kg/J	0.49	0.37
NO_x, kg/J	0.34	0.39
CO, kg/hr	0.69	3.29
Hydrocarbons, kg/hr	2.34	4.66
HCl, kg/hr	11.69	9.02
HF, kg/hr	0.84	0.61

Source: Data from [28]

6.3.4. Willow Island Generating Station, Allegheny Energy Supply Co., LLC.

The final case study is the Willow Island Generating Station Boiler #2, where TDF is fired along with woody biomass (see Chapter 4). This 188 MWe cyclone boiler is located in West Virginia, where major tire piles alone contain >6 million tires [31]. In 1993 the plant experimented with TDF, using hand feeding with 5 gal buckets. In 1995 the plant rented agricultural conveying equipment to further pursue TDF. In 1998 an extended test burn demonstrated the cost-effectiveness of this opportunity fuel. A project team was then formed to create a permanent system for receiving and handling the TDF. By the year 2000, TDF was being fired on a regular basis.

The TDF burned is nominally <37 mm (<1½") in particle size, and contains both belt and bead wire. The plant can burn up to 10 percent TDF, however typical firing rates are 5 – 6 percent TDF. In 2001, the plant burned 10,800 tonnes of TDF, representing over $1 x 10^6$ tires. The handling system is simple, receiving the TDF in an unused rotary car dumper and utilizing existing conveyor equipment (see Figure 6.7). This approach to using TDF as an opportunity fuel has proven both environmentally and economically advantageous [31].

6.4 Cofiring Tire-Derived Fuel in Stoker Boilers

Stoker boilers are another good match for TDF cofiring, especially traveling grate stokers [32]. The TDF chips are easily blended with the coal and completely combusted in the available time on the grate. Wire in the TDF is a concern as it may melt when the rubber has burned and potentially plug the grate keys. This section describes one successful demonstration of TDF cofiring in a stoker boiler.

6.4.1 Cofiring Tire-Derived Fuel at Jennison Generating Station, NYSEG

New York State Electric and Gas (NYSEG) burned approximately 2,721 tonnes (300,000 tires) of TDF [33]. These tests were conducted at NYSEG's Jennison Station located in Bainbridge, NY.

Figure 6.7. Receiving Tire-Derived Fuel at the Willow Island Generating Station.

Jennison station has an electrical generating capacity of 74 MW fired from four traveling chain grate stoker boilers consuming approximately 590 tonnes of coal daily. The station has the capability of burning the equivalent of 4,500,000 scrap tires annually, using a blend of 25% tire chips with 75% coal. This equates to an energy savings of approximately 59,000 tonnes of coal.

Cofiring TDF at Jennison Station required only minor changes in boiler operations. A noticeable amount of bead metal, and some steel belt material melted into nuggets that stuck on or in-between the stoker grate keys. Although the build up of metal was visible, it did not interfere with the ability to efficiently fire the blend. The ESP performance was not changed; there were no incidents of black smoke or increased emissions from coal/TDF blends. No reports of odor from tires burning occurred.

Sulfur dioxide emissions decreased somewhat when burning the TDF/coal blends. Other measured constituents changed only slightly depending upon the boiler load and TDF blends being used. In general, the emissions were about the same between the coal only and the coal/TDF blends.

The only differences noted in the chemical analyses of the bottom ash and fly ash samples were an increase in zinc noted in the fly ash and an increase in metal from the tire chips noticed in the bottom ash when firing the TDF blends compared to coal alone. Following these tests, magnetic separation equipment was installed at the ash pond to remove the ferrous metal from the bottom ash. This rendered the bottom ash acceptable for use by the local municipalities as a traction agent in the wintertime. A scrap metal dealer recycled the metal.

The industrial hygiene data collected demonstrated that there were no detrimental human, health or safety factors associated with the storage or handling of the tire chips at the plant. There were no noticeable odors from the tire chips, either inside or outside the plant, and the results of testing indicated there were no significant worker hazards associated with tire chip burning when compared to burning coal alone.

6.5. Cofiring TDF in Pulverized Coal Boilers

Pulverized coal (PC) boilers are the most common large combustion systems for the generation of electricity in the US and the industrialized economies of the world. PC boilers include both wall-fired boilers and tangentially-fired (T-fired) boilers; wall-fired boilers include both front-wall and opposed-wall configurations. Like cyclone boilers, they have the potential to use TDF as a fuel provided that a separate feed system he installed for the TDF chips. The high calorific value and low to medium sulfur content make it an attractive alternative fuel to be blended in the 5-10% (energy basis) range with coal.

6.5.1. General Characteristics

The constraints on pulverized coal cofiring of tire-derived fuel are more significant than with cyclone firing. The case study reported here is unique in that they utilized whole tries fed from the boiler roof.

Conventional TDF cannot easily pass through a pulverizer and must therefore be injected separately from the coal. Further, reduction of TDF to particle sizes appropriate for suspension firing is prohibitively expensive.

6.5.2. Cofiring TDF at FirstEnergy Toronto Plant

In the early 1990s, Ohio Edison completed a four-day scrap tire test burn of cofiring whole scrap tires and pulverized coal [34, 35]. Testing was done at the Toronto Plant, located in Toronto, Ohio. This facility is a 162 MW_n plant with three generating units still in service. TDF testing was done in Unit 5, a 42 MW_n B&W pulverized coal-fired wet slag bottom boiler. The boiler was modified to accept whole scrap tires.

The environmental results from the test burn showed that no permit requirements were violated. Comparing the baseline to the 20% tire (energy basis) cofiring test showed lead emission rates were 5% lower and particulate emissions rates were 28% lower. Based on the coal bunkered the last two days of the test at the 20% cofiring rate, the sulfur dioxide emission rate was 13.7% lower than the calculated expected value. Ohio Edison obtained a modification of its operating permit to include up to 20% of total boiler heat input from tires. At this rate, over three million scrap tires could be converted to electricity each year at the Toronto facility.

Figure 6.8 shows the emissions of NO_x, SO_2, and particulates during these TDF tests. The reported values are the average of three different measurements. The figure clearly shows more than a 35% reduction in NO_x when 20% TDF (energy basis) is used compared to the baseline. This may be due in part to the lower nitrogen content of the TDF fuel and the better distribution of heat in the boiler, reducing peak fireball temperatures. In addition there is relatively no change in the sulfur emissions or in the particulate emissions.

The mechanical results of the test burns showed the equipment performed according to design. Tire feed rates varied from one tire every 34 seconds to one tire every 10 seconds during the test. To prove the mechanical aspects of the delivery system a brief test was conducted with a feed rate of one tire every 7 seconds. However, with some minor mechanical changes and additional fine-tuning of the programmable logic

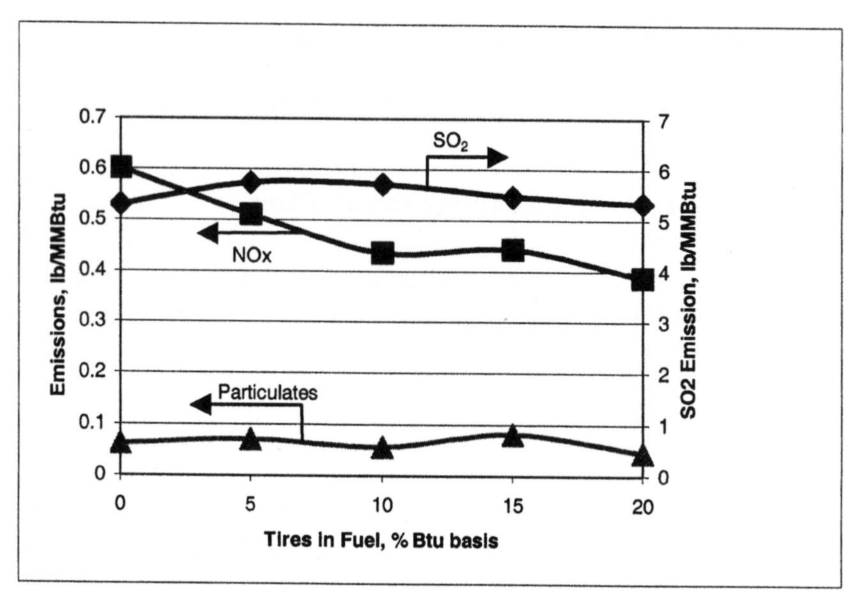

Figure 6.8. Emissions from the FirstEnergy TDF program.
Source: Data from [34]

controller for the delivery system, feed rates of one tire every four seconds are possible. A feed rate of one tire every four seconds corresponds to more than 40% energy input from tires. No boiler control problems were experienced and boiler slag tapping was accomplished without incident.

Based on the success of these tests, TDF has been evaluated at other FirstEnergy (Ohio Edison) stations – the Niles Station (cyclone-fired boiler) and the Burger Station (wall-fired boiler).

6.6. Cofiring TDF in Industrial Boilers

In addition to the examples of larger utility boiler demonstrations, there have been and continue to be many industrial-scale TDF cofiring programs [36 - 38]. Two examples are discussed below.

New Heights Power and Recovery facility in Ford Heights, IL can process about 3.7 million tires per year and is permitted to use up to 5 million tires per year in their electric generating unit [39].

International Paper's Lock Haven Mill currently burns a 15% TDF blend in their stoker boilers using about 9,000 tires daily eliminating about 3.2 million tires from landfill annually. The TDF has reduced the company fuel costs by about $200,000 per year [40].

6.7. Similar Opportunity Fuels – Waste Plastics

The disposal of municipal wastes is significant worldwide problem due to their partial biodegradability. Historically, these wastes have been landfilled similar to TDF. However, with the increase in municipal waste production combined with the decrease in available landfill sites, economic alternatives are being sought. One of these methods is combustion of municipal waste, especially the waste plastics. In 1997, there were 112 energy recovery facilities operating in 31 states throughout the United States with a designed capacity of nearly 101,500 tons per day [41].

6.7.1. Waste Plastic Composition

Waste plastics are composed primarily of low-density polyethylene (LDPE) and high-density polyethylene (HDPE) products. These two types make up nearly 60% of all plastic production. Figure 6.9 denotes the distribution of waste plastics by type [42]. Because plastics are made from hydrocarbons, they tend to have very high energy values as noted in Table 6.11. These values compare well with fuel oil and are considerably higher than most coals and TDF [42, 43].

6.7.2. Waste Plastic Utilization

Experts agree that properly equipped, operated and maintained incinerators or combustion facilities can meet the latest U.S. emissions standards while cofiring waste plastics [43]. In fact, plastics can be successfully burned in dedicated energy recovery facilities that achieve

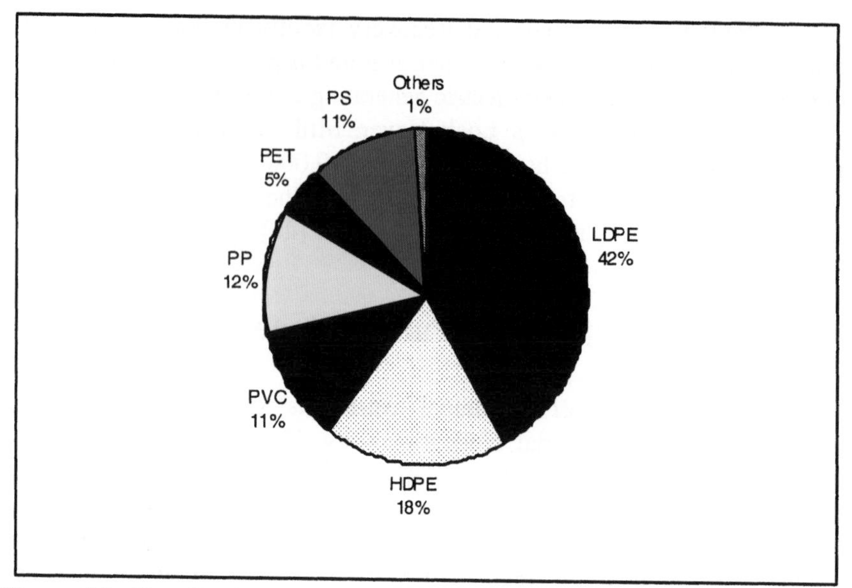

Figure 6.9. Distribution of Waste Plastics by Type
Source: Data from [42]

high combustion temperatures (>1100°C or 2012°F) to eliminate dioxin and furan production. One concern in utilizing waste plastics is in preventing melting of the material rather than combusting the material. If a melted plastic reaches a tube surface, it could stick and reduce heat

Table 6.11. Heating Value of Some Common Plastics

Material	Heating Value	
	kJ/kg	Btu/lb
High Density Polyethylene (HDPE)	46,300	19,900
Polyvinyl Chloride (PVC)	17,500	7,500
PET	22,700	9,800
Low-Density Polyethylene (LDPE)	46,300	19,900
Polyethylene (PE)	46,290	19,900
Polypropylene (PP)	46,170	19,850
Polystyrene (PS)	41,400	17,800
Fuel Oil	48,600	20,900

(Data from [40, 41])

transfer to the steam. In addition, the sticky surface would permit solid ash particles to agglomerate and increase deposition. Another concern is the potentially high chlorine content of the plastic wastes. Due to corrosion potential, this may limit the amount of plastics that can be used in a blend.

6.8. Conclusions Regarding TDF and Related Opportunity Fuels

Several conclusions can be drawn regarding the use of TDF cofired with coal in industrial and utility boilers. These include:

- There is a sufficient supply of scrap tires in most areas of the US to warrant the investigation of TDF cofiring in boilers.
- ASTM has recently released a standard procedure for processing TDF; this will aid users in determining proper specifications for their individual units.
- The boilers best suited for using TDF are cyclones, stokers, and fluidized bed boilers. Some unique situations exist with TDF in pulverized coal boilers. These boilers can use up to 20-25% TDF (energy basis).
- Using tire chips greater in size than about 25 mm in cyclone boilers has resulted in plugging hammer mills, coal transfer chutes, etc. even in small cofiring amounts. Therefore, the TDF should be introduced after the hammer mills as the coal is proceeding directly to the bunkers.
- Steel in the tire chips (from the bead and steel belt) can cause problems in the handling of the TDF and in some instances in the burning of the fuel.
- For best results in obtaining a consistent TDF feed material, most utilities have gone to chipping the tires themselves. They can then control the size of the chip and the consistency of the product.
- TDF blends did not result in any major boiler operating problems. Full-load was achievable at reasonable excess air

levels. Unburned chips in the bottom slag have presented a problem if the chips are too large. Steel strings in the ash/slag have resulted in the need for a magnetic separator to remove them prior to selling the product.

- Boiler efficiencies were equal to or slightly greater with the TDF blends due to its higher calorific value and low moisture content.
- Gaseous emissions do not appear to be negatively impacted unless a low sulfur coal is being blended with the TDF, then the SO_2 emissions might rise slightly. NO_x emissions tend to be reduced with TDF cofiring and opacity appears unchanged or slightly lower with the TDF. CO and hydrocarbon emissions are increased when cofiring due to insufficient time to completely burn out large particles. This can be mitigated to some extent by limiting the size of the tire chip.
- Zinc emissions are higher with the TDF blends, but not above the regulated limits.
- Ash leaching tests did not show any difference between baseline and TDF cofiring tests.
- Waste plastics and similar materials with high heating values can be successfully cofired for energy recovery.
- Utilization of waste plastics must be done in facilities that produce a high enough temperature to completely combust the plastic to prevent deposition and dioxin and furan production.

6.9. References

1. McGowin, C.R., 1991. "Alternate Fuel Cofiring with Coal in Utility Boilers," EPRI Proceedings: 1991 Conference on Waste Tires as a Utility Fuel, EPRI GS-7538.

2. Winslow, J., Ekmann, J., Smouse, S., Ramezan, M. and Harding, N.S., 1996. "Cofiring of Coal and Waste," International Energy Association Report, IEACR/90.
3. Harding, N. S., 2002. "Cofiring Tire-Derived Fuel With Coal," 27th International Technical Conference on Coal Utilization and Fuel Systems, Clearwater, FL.
4. Niemeyer, S., 1977. "Managing Wastes: Tires," Nebraska Cooperative Extension NF94-197.
5. "Waste Tire Perspective," 1997. Louisiana Department of Environmental Quality.
6. Environmental Waste International, 2001. "Scrap Tire Supply," Advanced Technology for Specialty Waste Streams.
7. Barnhart, R.R., 1985. "Rubber Compounding," in Concise Encyclopedia of Chemical Technology, Kirk-Othmer, eds., John Wiley & Sons, Inc.
8. Sax and Lewis, 1997. Hawley's Condensed Chemical Dictionary, Thirteenth Edition, John Wiley & Sons, Inc., 1997.
9. McGrath, J.F. 1985. "Elastomers, Synthetic" in Concise Encyclopedia of Chemical Technology, Kirk-Othmer, eds., John Wiley & Sons, Inc., Pp. 391.
10. Bauer, R.G. "Styrene-Butadiene Rubber" in Concise Encyclopedia of Chemical Technology, Kirk-Othmer, eds., John Wiley & Sons, Inc., Pp. 399-400.
11. Dannenberg, 1985. in Concise Encyclopedia of Chemical Technology, Kirk-Othmer, eds., John Wiley & Sons, Inc.
12. Skolnik, L., 1985. "Tire Cords," in Concise Encyclopedia of Chemical Technology, Kirk-Othmer, eds., John Wiley & Sons, Inc.
13. Goodyear Tire and Rubber Company, "Scrap Tire Recovery," 2002.
14. Koziar, P. J., 1991. "Overview of Regional Waste Tire Management Opportunities for Electric Utilities," EPRI Proceedings: 1991 Conference on Waste Tires as a Utility Fuel, EPRI GS-7538.
15. U.S. Scrap Tire Market, 2002. Scrap Tire News Reports, Scrap Tire Management Council.
16. ASTM Document Summary, 2001. "Practice D6700-01 Standard Practice for Use of Scrap Tire-Derived Fuel," ASTM, West Conshohocken, PA.
17. MSW Management News, 2001. "Tire-Derived Fuel Standard Released by ASTM."
18. American Recycler, 2002. "Equipment Spotlight – Tire Shredders."

19. Bakkom, T. K. and M. R. Felker, 1991. "Tire Shredding Equipment," EPRI Proceedings: 1991 Conference on Waste Tires as a Utility Fuel, EPRI GS-7538.
20. Granger, J. E. and G. A. Clark, 1991. "Fuel Characterization of Coal/Shredded Tire Blends," EPRI Proceedings: 1991 Conference on Waste Tires as a Utility Fuel, EPRI GS-7538.
21. Reisman, J.I, and Lemieux, P.M., 1997. "Air Emissions from Scrap Tire Combustion," EPA Report No. EPA-600/R-97-115.
22. Tillman, D. A. 1994. Trace Metals in Combustion Systems, Academic Press, San Diego, CA. Pp. 140-141.
23. Tillman, D.A., 1996. "Report of the Cofiring Combustion Testing at the Allen Fossil Plant Using Utah Bituminous Coal as the Base Fuel," Final Report to Tennessee Valley Authority and Electric Power Research Institute, prepared by Foster Wheeler Environmental Corporation.
24. Weinhold, J.F., 1993. "TDF Cofiring Tests In A Cyclone Boiler," Strategic Benefits of Biomass and Waste Fuels, Electric Power Research Institute, Washington, D.C.
25. Costello, P. A., R. G. Waldron, and W. H. Witts, 1996. "Tire Derived Fuel and Thermal Treatment Waste Incineration – Commercial Operation in Coal Fired Cyclone Units," ASME FACT Division, 1996 Joint Power Generation Conference, ASME.
26. Stopek, D.J., et. al., 1991. "Testing of Tire-Derived Fuel in a Cyclone-Fired Utility Boiler," EPRI Proceedings: 1991 Conference on Waste Tires as a Utility Fuel, EPRI GS-7538.
27. Stopek, D.J., Licklider, P.L, Millis, A.K, and Diewald, D.J., 1993. "Tire Derived Fuel (TDF) Cofiring in a Cyclone Boiler – At Baldwin Station," Strategic Benefits of Biomass and Waste Fuels, Electric Power Research Institute, Washington, D.C.
28. Nast, V., Eirschele, G. and Hutchinson, W., 1993. "TDF Co-firing Experience In A Cyclone Boiler," Strategic Benefits of Biomass and Waste Fuels, Electric Power Research Institute, Washington, D.C.
29. Hutchinson, W., et. al., 1991. "Experience with Tire-Derive Fuel in a Cyclone-fired Utility Boiler," EPRI Proceedings: 1991 Conference on Waste Tires as a Utility Fuel, EPRI GS-7538.
30. Letheby, K., 2002. "Utility Perspectives on Opportunity Fuels," 27th International Technical Conference on Coal Utilization and Fuel Systems, Clearwater, FL.

31. Holt, G. 2003. Alternate Fuels. Presented at the Community Advisory Panel Meeting, Peasants/Willow Island District, West Virginia. July 7.
32. Harding, N.S. and Owens, W.D, 1994. "The Utilization of Waste Tire-Derived and Railroad-Tie-Derived Fuels in Coal-fired Stoker Boilers," 10th Annual Coal Preparation, Utilization and Environmental Control Contractors Conference, Pittsburgh, PA.
33. Murphy, P.M. and Tesla, M.R., 1993. "Co-firing Tire Derived Fuel In A Stoker Fired Boiler," Strategic Benefits of Biomass and Waste Fuels, Electric Power Research Institute, Washington, D.C.
34. Gillen, J.E. and Szempruch, A.J., 1993. "Ohio Edison Tires-to-energy Project," Strategic Benefits of Biomass and Waste Fuels, Electric Power Research Institute, Washington, D.C.
35. Horvath, M., 1991. "Results of the Ohio Edison Whole-tire Burn Test," EPRI Proceedings: 1991 Conference on Waste Tires as a Utility Fuel, EPRI GS-7538.
36. "Northern States Power's Interest in Biomass Energy Continues to Grow," United BioEnergy Commercialization Association Bulletin, Autumn 1994.
37. "MU Power Plant," Energy Management Power Plant, University of Missouri Columbia, June 2000.
38. Maley, S.M. 2003. Production of Tire Derived Fuel: Use of Mobile System as a Viable Economic Alternative. Proc. 28th International Technical Conference on Coal Utilization and Fuel Systems, Clearwater, FL. March 9 – 13.
39. McMurray, D., 2000. "Facilities Should be Adequate to Process Recalled Tires in Illinois," Illinois Environmental Protection Agency.
40. Brennan, J., 1998. "1998 Governor's Awards for Environmental Excellence," International Paper Company, Lock Haven Mill, Pennsylvania Department of State.
41. Plastic Resources, "Waste-to-Energy," 1999.
42. Eulalio, A. C., N. J. Capiati and S. E. Barbosa, 1999. "Municipal Plastic Waste: Alternatives for Recycling with Profit."
43. Piasecki, B. D. Rainey and K. Fletcher, 1998. "Is Combustion of Plastics Desirable?" American Scientist, July-August.

CHAPTER 7: GASEOUS AND LIQUID OPPORTUNITY FUELS

7.1. Introduction

For the most part, opportunity fuels of primary significance to the electric utility and process industry communities are solids—as is reflected in the previous chapters. However numerous gaseous and liquid opportunity fuels also exist. These fuels are typically methane-rich gases such as (not exhaustive) methane recovered in association with coal mining, off-specification refinery gas, coke oven gas, landfill gas, and wastewater treatment gas. Liquid opportunity fuels include hazardous wastes and waste oils that may or may not be considered as hazardous wastes. These fuels are used mainly in small quantities as blends with other fossil fuels or in specialty niche markets. This chapter discusses gaseous and liquid opportunity fuels such as coalbed methane, landfill gas, coke oven gas, and wastewater treatment gas as well as hazardous liquids and waste oils used in cement kiln and other energy applications.

While there is no good estimate of the quantities of these opportunity fuels, mainly because they continue to be produced by other processes, they are a potentially significant source of energy in many applications. In all probability, the gaseous opportunity fuels are produced in greater quantity than the liquid opportunity fuels. Both are treated in this chapter.

7.1.1. Common Characteristics of the Gaseous Opportunity Fuels

The common thread unifying the gaseous opportunity fuels is the relatively high concentration of methane gas. Methane is the desirable component in coal bed gases, if these gases are extracted for use. Methane is also the desirable component in landfill and wastewater treatment gas, and is a significant component in refinery off-gas.

Analysis of gases trapped in ice indicates that the amount of atmospheric methane has more than doubled over the past three centuries and that it has increased at the rate of 1 percent per year during the past 15 years [1]. This increase is correlative with the growth of human population and increase in human activity. Natural systems such as wetlands and decomposing forested areas account for about 40 percent of the methane released to the atmosphere. Significantly, the balance is largely the result of human activities such as rice cultivation (19 percent), livestock (11.5 percent), landfills (8 percent), biomass burning (11.5 percent), venting from oil and gas wells (4 percent), and coal mining (6 percent). Most experts agree that little reduction can be gained from altering livestock and rice cultivation methods; however, some important reductions can be achieved by improving techniques for recovery and utilization of methane from coal mines.

Selected fundamental properties of methane are shown in Table 7.1. These properties are well understood and well defined in the literature.

Tab le 7.1. Fundamental Properties of Methane

Property	Value
Molecular weight	16.04
Melting point, °C (°F)	-182.3 (-296)
Boiling point, °C (°F)	-162 (-259)
Explosivity limits (Vol %)	5.3 – 14.0
Autoignition temperature, °C (°F)	538 (1001)
Heat of combustion (kJ/mol)	882
Heat of combustion, MJ/m^3 (Btu/ft^3)	37.76 (1013)
Heat of combustion MJ/kg (Btu/lb)	55.41 (23,875)

Sources: [2, 3]

Additionally, the kinetics of methane pyrolysis and oxidation have been well defined. Dellinger et. al. [4] have placed the temperature at which methane begins to pyrolyze into the radicals and fragments, $CH_3\bullet$ and $H\bullet$, at 660°C (1,220°F) and gives the following Arrhenius equation terms for methane pyrolysis: pre-exponential constant (A) = 3.5×10^9 (1/sec) and the activation energy, E = 200.9 (kJ/mol). Dellinger et. al. [4] report methane oxidation kinetic parameters as determined by Union Carbide to be as follows: A = 1.68x1011 (1/sec) and E = 218.1 (kJ/mol). These basic parameters govern the behavior of most waste gases used as opportunity fuels. The primary gases are reviewed below.

7.2. Coalbed Methane

The coalification process, whereby plant material is progressively converted to coal, generates large quantities of methane-rich gas that are stored within the coal. The presence of this gas has been long recognized because of explosions and outbursts associated with underground coal mining. Only recently has coal been recognized as a reservoir rock as well as a source rock, thus representing an enormous underdeveloped "unconventional" energy resource [1]. But production of coalbed methane is accompanied by significant environmental challenges, including prevention of unintended loss of methane to the atmosphere during underground mining, and disposal of large quantities of water, sometimes saline, that are unavoidably produced with the gas.

Most gas in coal is stored on the internal surfaces or organic matter. Because of its large internal surface area, coal stores 6-7 times more gas than the equivalent rock volume of a conventional gas reservoir [1]. Gas content generally increases with coal rank, with depth of the coalbed, and with reservoir pressure. Fractures or cleats that permeate coalbeds are usually filled with water; the deeper the coalbed, the less water is present, but the more saline it becomes. In order for gas to be released from the coal, its partial pressure must be reduced, and this is accomplished by removing water from the coal bed. Large amounts of water are produced from coalbed methane wells, especially in the early stages of production. While economic quantities of methane can be produced, water disposal options that are environmentally acceptable and yet economically feasible become a concern.

7.2.1 Coalbed Methane Resources

Reducing the internal pressure during coal mining results in the release of coalbed methane. Unfortunately, most of this gas is presently emitted to the atmosphere through mine ventilation systems and by mine degasification systems as a safety measure. The in-place coalbed methane resources of the United States are estimated to be more than $19.8x10^9$ m^3 ($700x10^9$ ft^3), but less than $2.8x10^9$ m^3 ($100x10^9$ ft^3) may be economically recoverable. Worldwide estimates of in-place resources are as much as $212x10^9$ m^3 ($7.5x10^{12}$ ft^3), but this number is uncertain because of the scarcity of basic data on coal resources and gas content. On a tonnage basis, emissions from the world coal industry can be estimated at 25 million tonnes per year of which nearly 24 million tonnes are discharged into the atmosphere. This atmospheric discharge equates to about 4-6% of the global methane emissions [1, 5 - 6]. As a result, coalbed methane gas is now one of "the gases to be subject to mandatory cutbacks" by the Kyoto protocol [7].

More than 5% of atmospheric methane resulting from human activity is derived from coal mining, and fully one-third of this amount is derived from underground coal mines in China. The IEA published a comprehensive report on the methane emissions from coal industry in the ten largest coal-producing countries; Table 7.2 summarizes these data [5].

7.2.2. Coalbed Methane Deposits – United States

Coalbed methane can be used as an energy source that is environmentally more acceptable than mining and combustion of coal. It can partly replace coal as a fossil energy source, and it sometimes occurs where other conventional resources of oil and gas are not present. Coalbed methane accumulations are widespread, commonly basinwide, and are characterized by large in-place resources. Although most wells will encounter gas in these widespread accumulations, production rates will be highly variable, even within a small area, because of the heterogeneous nature of coalbeds. Basin-wide studies are needed to determine controls of the occurrence, availability, and recoverability of coalbed methane in the United States and other countries that need clean

Table 7.2. Methane Emissions from the Ten Largest Coal Producing Countries

Country	1990 Coal Production (10^6 tonnes)	Coal Bed Methane Production (10^6 tonnes)		
		Gross Production	Total Utilized	Net Methane Emissions
China	1,053	7.70	0.14	7.56
Former USSR	703	5.02	0.19	4.83
USA	931	3.61	0.31	3.30
Germany	434	1.22	0.25	0.97
Poland	216	1.35	0.14	1.21
UK	95	0.86	0.10	0.76
Australia	163	0.45	0.08	0.37
South Africa	206	0.85	---	0.85
India	212	0.45	---	0.45
Czech Republic	119	0.39	0.09	0.30
TOTAL	4,132	21.90	1.30	20.60
TOTAL WORLD	4,704	25.0*		

Note: *Assuming the same relationship between methane emissions and coal production at the coal producing countries, the worldwide methane emission from coal production would be approximately $25x10^6$ tonnes.

Source: [5]

energy resources. Underground coal-mining areas, such as the Appalachian basin, should be emphasized because of the need to reduce atmospheric methane emissions. Most previous exploration and research efforts have been in the San Juan basin and the Black Warrior basin. However, since each coal-bearing basin has unique attributes, coalbed methane issues need to be studied separately in each basin. Figure 7.1 shows the locations of the largest coalbed methane deposits in the United States. This corresponds to the major coalfields in the US.

Studies by the USEPA document that the mines with the most prominent methane resources are in West Virginia, Pennsylvania, Virginia, Alabama, and Colorado [8]. Their studies identify 21 gassy

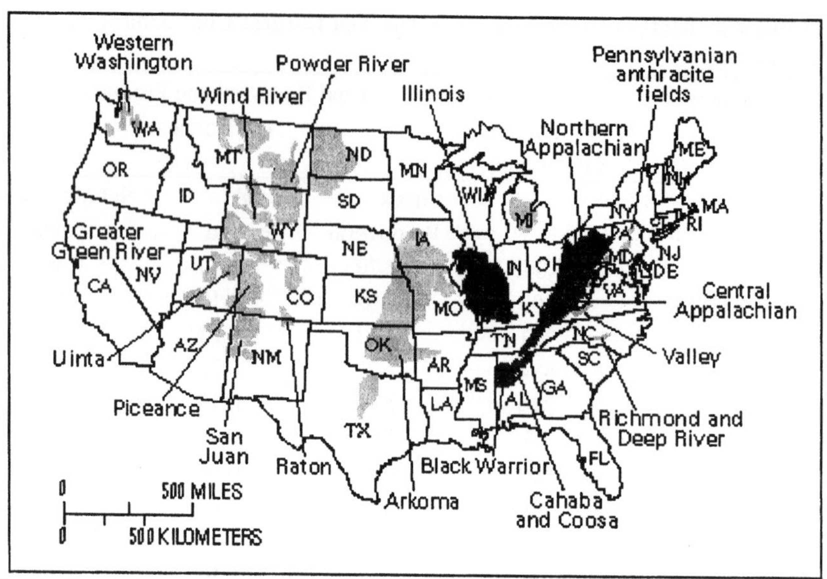

Figure 7.1. Coalbed methane deposits within the United States.
Source: [1]

mines with a cumulative production of 4.8×10^6 m3/day (170×10^6 ft^3/day) or 176×10^6 m^3/yr (62×10^9 ft^3/yr). This is equivalent to 180 TJ/day (170×10^9 Btu/day) or 6.5 PJ/yr (62×10^{12} Btu/yr).

7.2.3 Coalbed Methane Extraction and Utilization Technologies

For coalbed methane, there are fundamental differences between the methods for the methane gas extraction. During coal mining operations, the methane gas is eliminated from the mining area by the mine ventilation system or through the de-gasification pipelines under the coal mine safety regulation. The methane emissions from the mined out area can be discontinued by deluging the area, or, if left alone, it can be expected to supply a steady source of coalbed methane gas if the area is kept under proper conditions.

If no mining is occurring, the coalbed methane can be extracted by drilling several boreholes and pumping it out. This is a popular operation in Australia [9]; this method is also used to provide the supply of city gas and/or chemical feedstock through a pipeline to the market. In this case, the coalbed methane gas will continue to be used and not vented, thereby reducing any effect on global warming.

The majority of the coalbed methane is released to the atmosphere through the mine ventilation system as a dilute fugitive gas. About 80 percent of the methane released by coal mining is exhausted as atmospheric emissions. The remaining 20 percent of the coalbed methane gas is recovered through the boreholes and pipelines at higher methane concentrations for use. The goal is to increase the recovery rate of the high-grade useful coalbed methane gas and decrease the amount escaping as a fugitive greenhouse gas. Table 7.3 presents some options for reducing the amount of fugitive coalbed methane gas in order to make it more available for use.

As is apparent from Table 7.3, the stage of mining and the approach to coal bed methane has a significant impact upon the utilization strategy for this opportunity fuel. Methane extracted before mining, or through an integrated methane extraction/coal mining operation, can be used in a highly flexible manner. It can be moderately conditioned to remove any moisture or other contaminants and compressed for insertion into natural gas pipelines. Alternatively it can be used as electricity generation fuel in boilers or combustion turbine applications. It can be used as a feedstock for liquid fuels and chemicals. Methane extracted at sufficient concentrations to approach a synthesis gas (e.g., 11 – 29 MJ/m^3 or 295 – 780 Btu/ft^3) can be used in reasonable proximity to the coal mine. At the higher values, it can be compressed and injected into a pipeline if it is a small fraction of the flow of natural gas in that pipeline.

Methane extracted during mining, at very low concentrations, is more difficult to use. However experiments are underway to develop technologies for using this methane as an opportunity fuel on site. If a power plant is immediately adjacent to the mine, the methane extracted in an air stream can be used as combustion air in a boiler or combustion turbine. The determining factor is the location of the mine relative to the generating equipment—and the consequent cost of ductwork and associated equipment.

Methane extraction and utilization is well practiced in the USA and around the world. The CONSOL Buchanan No 1 mine, along with VP No. 3 and VP No. 8, combined, produce and productively utilize $2.07x10^6$ m^3/day (73x106 ft^3/day) of coalbed methane [8]. These mines

Table 7.3. Options for Methane Recovery from Coal Mines

	Mining Activity			
	Pre-mining degasification	During mining vent air degas	Enhanced gob well recovery	Integrated CH_4 recovery and use
Recovery Techniques	Vertical wells; in-mine boreholes	In-mine boreholes; vertical wells & fans	Vertical gob wells; in-mine boreholes	All techniques
Support technologies	In-mine drills and/or advanced surface rigs, compressors, and other support facilities	In-mine drills and/or basic surface rig, surface fans, and ducting heat recovery from vent gas	In-mine drills and/or basic surface rig, surface fans, compressors, system pumps, and other support facilities	All technologies; ability to optimize degasification using combined strategies
Expected gas quality	High – 32-37 MJ/m^3 (860 – 1000 Btu/ft^3); > 90% CH_4	Low; - <1% CH_4	Medium – 11-29 MJ/m^3 (295 – 780 Btu/ft^3); 30 – 80% CH_4	All qualities, especially high quality gas
Methane reductions	>70%	10 – 90% recovery	>50%	80 – 90%
Gas Use Options	Chemical feedstocks; power generation; pipeline gas distribution	Combustion air for on-site power generation or other combustion uses	On-site power generation; co-gas distribution; industrial use	Chemical feedstocks; power generation; pipeline gas; liquid fuel production (LNG, MeOH)
Technology availability	Currently available	Demonstration required for vent gas	Currently available	Currently available

Source: [6]

are located in Buchanan County, Virginia. US Steel uses methane from its Oak Grove and Pinnacle No. 50 mines in West Virginia. JWR Mining produces and uses 1.13×10^6 m^3/day (40×10^6 ft^3/day) at Blue Creek No. 3 – No 7 mines in Tuscaloosa County, Alabama. These are significant uses of coalbed methane.

Perhaps the largest use of coalbed methane is at the Appin and Tower coal mines in New South Wales, Australia. There coalbed methane is extracted from boreholes advancing ahead of the mining operations. The methane is then ducted to internal combustion engines generating 94 MW_e (gross). Ventilation air with minor concentrations of methane is used for combustion air in the IC engine-generator sets. The electricity so generated is both used on-site and sold to the local utility.

Another innovative approach to coalbed methane utilization as an opportunity fuel is found in the Ukraine and in the Czech Republic. There the coalbed methane is recovered, compressed, and used as fleet vehicle fuel [9]. This also resolves the issue of proximity of the fuel to the point of use.

In summary, coalbed methane represents a significant potential source of opportunity fuel. The key is matching the resource to an end use either by proximity to a natural gas pipeline, or by installation of electricity generating equipment or related energy systems at or near the point of coalbed methane generation.

7.3 Landfill Gas

Landfill gas is produced by decomposition of waste materials collected in municipal waste areas. Each person in the United States generates about 2.0 kg (4.5 lbs) of waste per day, or almost 750 kg (1700 lbs) per year, most of which is deposited in municipal sold waste (MSW) landfills [11-12]. Table 7.4 lists the composition of municipal waste for the year 2000 [11].

Most cities have a landfill where the population may discard their unwanted trash. Major metropolitan areas such as Los Angeles have numerous landfills. In the case of Los Angeles County, landfills may become very large. Puente Hills, for example, accepts upwards of 10 million tonnes of material annually. Regional landfills may also be built to serve numerous cities and counties. Roosevelt Regional Landfill in

Table 7.4. Estimated Municipal Waste Composition for 2000

Component	Weight %
Paper and Paperboard	41.0
Yard Wastes	15.3
Food Wastes	6.8
Plastics	9.8
Wood	3.8
Textiles	2.2
Rubber and Leather	2.4
Glass	7.6
Metals	9.0
Miscellaneous	2.1
TOTAL	100

Source: [9]

south central Washington State, for example, accepts some 3.5 million tonnes/yr of waste from Everett and Tacoma, Washington plus communities as far south as northern California.

7.3.1 *Landfill Gas Composition*

As MSW decomposes, it produces a blend of several gases, including methane (about 50%). Table 7.5 shows the main constituents of landfill gas and their proportions.

Table 7.5 Landfill Gas Constituents

Constituent Gas	Concentration in Landfill Gas	
	Range	Average
Methane (CH_4)	35-60%	50%
Carbon Dioxide (CO_2)	35-55%	45%
Nitrogen (N_2)	0-20%	5%
Oxygen (O_2)	0-2.5%	<1%
Hydrogen Sulfide (H_2S)	1-1,700 ppm_v	21 ppm_v
Halides	NA	132 ppm_v
Water Vapor (H_2O)[1]	1-10%	NA
Nonmethane Organic Cmpds.	237-14,294 ppm_v	2,700 ppm_v

Note: [1]Landfill gas, as extracted, is saturated with water vapor

Sources: [11-12]

The methane in landfill gas not only is a greenhouse gas, but also poses explosion hazards if uncontrolled. On the other hand, it is the main component of natural gas and can be a valuable source of energy. The carbon dioxide in landfill gas is also a greenhouse gas. The non-methane hydrocarbons are a collection of trace quantities of numerous haloginated and non-haloginated compounds; the actual concentrations depend upon the waste being decomposed and the conditions in the landfill.

The landfill is, in reality, a large bioreactor. Each landfill cell subjects the waste interred therein to temperature and pressure. Further, depending upon the climate and the operating parameters, the landfill cell subjects the waste to varying concentrations of moisture. Initially the municipal waste decomposes using processes of aerobic decomposition, consuming the air deposited with the municipal waste. This yields a gas rich in CO_2. As the air is consumed, the mechanism of degradation converts to anaerobic decomposition; it is the anaerobic decomposition that produces CH_4. Landfills also produce leachate—a liquid product containing both water and products of decomposition. Some landfills are permitted to reinject the leachate into the landfill, thereby enhancing CH_4 production. This practice both increases CH_4 production and also accomplishes this production earlier in the time line of the landfill.

7.3.2 *Landfill Gas Energy Utilization Options*

Landfill gas utilization for energy production has increased in recent years, as is shown in Figure 7.2. Some 60 percent of the electricity generated from MSW comes from landfill gas [14]. Such energy recovery can come from large or small operations using a wide variety of technologies. It can be estimated that, in the year 2000, landfill gases accounted for nearly 1.1×10^9 kWh of electricity generated in the USA [13-14]. Landfill gas has been used as an opportunity fuel for non-electricity options as well.

Landfill gas usage has been promoted by utility deregulation and the marketing of "Green Power." Landfill gas has been accepted as a source of "green energy" by The Center for Resource Solutions, the major private organization certifying green power programs. Further, it has been recognized as a biomass energy source in virtually all proposed energy legislation at the Federal level of the USA.

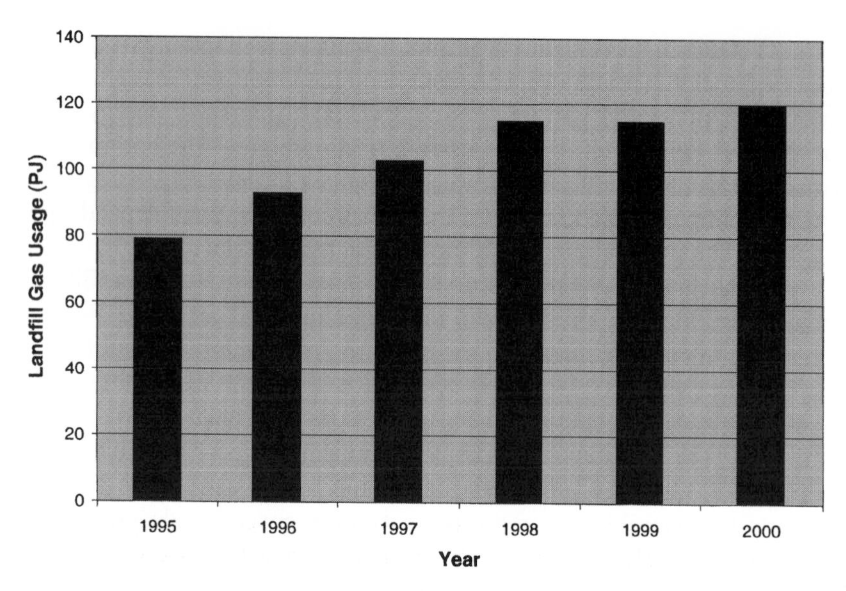

Figure 7.2. Recent Growth in US Utilization of Landfill Gas as an Opportunity Fuel.
Source: [13]

There are five main ways to recover energy from landfill gas [15]: direct heating, electricity generation, chemical feedstock, purification to pipeline-quality gas, and heat recovery, as shown in Table 7.6. Each of these methods has a variety of landfill gas applications. The most common method for energy recovery involves dehydrating landfill gas and injecting it directly into reciprocating internal combustion engines. For very large facilities, such as Puente Hills landfill, steam boilers are more appropriate. The Puente Hills landfill has a 50 MWe boiler for electricity generation. That landfill has also installed internal combustion engines, a combustion turbine, microturbines, and fuel cells to evaluate all energy recovery options. It also has installed and operated facilities to purify and compress landfill gas for use as fleet vehicle fuel.

The advantages of energy recovery include cheap fuel plus decreased landfill emissions of methane, non-methane organic carbon compounds (NMOCs), and toxics (e.g., benzene, carbon tetrachloride, and chloroform). Although carbon dioxide emissions increase with the energy recovery option, the net atmospheric balance is a positive one

Table 7.6. Landfill Gas Energy Recovery Options

Application	Option
Direct Heating Options	Use in industrial boilers
	Use in space heaters
	Use in industrial heating systems
Electricity Generation Options	Internal combustion engines
	Combustion turbines and microturbines
	Combined cycle plants
	Steam boilers and steam turbines
	Fuel cells
Vehicle Fuel Options	Purification and compression
	Conversion to methanol
	Conversion to diesel fuel (Fischer-Tropsch)
Pipeline Quality Gas	Use as vehicle fuel
	Incorporation into local natural gas pipeline
Heat Recovery from Flares	Use in Organic Rankine Cycle Systems
	Use in Stirling Cycle Engines

because carbon dioxide emissions are significantly less radiative (i.e., the alleged "greenhouse effect" is less) than methane emissions.

7.3.3 Landfill Gas Recovery Economics

The most economical options for landfill gas utilization are direct uses such as process heat and boiler fuel, where the end users are in close proximity (no more than1.6-3.2 km [1-2 miles]) from the landfill, and whose gas supply needs closely match production at the landfill. In practice, end users are infrequently located near landfills and rarely require continuous fuel in the amounts produced [16]. As of 1992, there were 21 landfills (less than 20% of total energy recovery projects) with use of landfill gas as heating fuel [17].

Boiler fuel is the most typical direct use and a particularly attractive option since conventional equipment can be used with minimal modifications. Boilers are generally less sensitive to moisture in the gas, or landfill gas trace constituents, and therefore require less cleanup than other alternatives. End-use options include industrial applications such as kilns, lumber drying, oil refining, hotel heating, and cement manufacturing. These tend to be economical applications because of the continuous need and availability of the fuel. Another direct approach involves modest gas upgrading followed by gas compression and utilization as a vehicle fuel—particularly for off-road vehicles.

Electricity generation is also frequently economically attractive to pursue. Generally, there are three applications for landfill gas electricity generation: internal combustion (IC) engines, gas turbines, and fuel cells. As of 1992 there were about 61 projects that generated electricity using IC engines and 24 using turbines, accounting for a total output of 344 MW. IC engines are most economical in the 1-3 MW range, while turbines are more economical greater than 3 MW [17-19].

In the future, fuel cells may become attractive because of their higher energy efficiency, negligible emissions impact, and suitability for all landfill sizes. Some studies suggest that fuel cells would be more competitive in small (<1 MW) to medium (<3 MW) projects [15]. According to a study by the Electric Power Research Institute, if individual fuel cell power plants were used at landfills, 6,000 MW of electricity could be generated from landfill gas [20]. Another study prepared for the EPA suggests that the approximate total power output that could be generated from about 7,500 landfills using fuel cell energy recovery could be 4,370 MW [15].

The use of landfill gas as a chemical feedstock is less economically attractive. This option involves the use of expensive clean-up, purification, and processing equipment to bring the landfill gas to the quality standards of common chemical feedstocks, such as natural gas. Using landfill gas as a chemical feedstock remains largely uneconomical as long as the price of conventional feedstocks remains low. Other disadvantages are high transportation costs and a need for proximity to the end user. Landfill sites have found that gas pipelines cannot exceed 1-2 miles to be const-effective [17].

Gas purification to pipeline quality is yet another option. This option involves the conversion of landfill gas, a medium heating value gas, into high heating value gas for local gas distribution networks or, in compressed form, for vehicular fuel. In 1992, there were 7 sites that upgraded landfill gas to pipeline-quality gas. This option also remains uneconomical as long as the prices of natural gas and fuel oil remain relatively low [16].

7.3.4. *Development of the Landfill Gas Industry*

The first commercial gas energy recovery project was at the Palos Verdes Landfill, in Rolling Hills, CA in 1975. The project converted landfill gas to pipeline-quality gas that was sold to the southern California Gas Company. Several other projects to convert landfill gas to pipeline-quality gas were started in the late 1970s in California, including Mountain View in 1978 and Monterey Park in 1979. The first direct heating boiler projects were brought on line in the late 1970s and early 1980s. One of the first electricity generation projects took place at Brattleboro, VT in 1982. The largest project is the Puente Hills boiler; Los Angeles County Sanitation District also has installed the 12 MW_e Spadra boiler and a companion project, the Coyote Canyon Boiler. Most projects are located in California and the northeast.

The average size of a landfill gas energy recovery project is about 3 MW, with typically over 95% availability. The number of commercial landfill gas energy recovery projects has grown from 4 in 1981 to about 130 in 1996. Even though there has been a large increase in projects, EPA estimates that over 700 landfills across the United States could install economically viable landfill gas energy recovery systems.

Unfortunately, about 30 of the original conversion and direct use projects initiated in the 1970s and 1980s have had to shut down due to more competitive market conditions of the 1990s. Therefore, although the advantages of landfill gas energy recovery are many, there are few successful commercial projects relative to the number of landfills due to prevailing market conditions and the array of other formidable barriers that confront project developers. Some of these barriers include low oil and gas prices; the need for expensive new, sometimes untested, technology; high transportation costs for landfill gas (e.g., dedicated pipelines for relatively small supplies); high debt-service rates for projects that generate electricity or pipeline-quality gas; a limited or unstable marketplace; obtaining third-party project financing at reasonable cost (financing is difficult, time-consuming, and proportionately more costly for small projects than for large ones); difficulties obtaining air permits, especially for projects located in ozone, nitrogen oxide, and carbon monoxide nonattainment areas; difficulties in negotiating power contracts with local utilities because they are primarily interested in purchasing low-cost power without considering environmental externalities; unforeseen costs resulting from compliance with new air quality rules and regulations, and declining energy revenues that cannot be adjusted to offset new costs; taxation by some States (e.g., California) on landfill gas extraction and energy conversion facilities; and difficulties in complying with overlapping Federal and State energy policies and environmental regulations.

Of these barriers, the most significant is low oil and natural gas prices, which make recovery and conversion, with its high initial capital costs, lack of economies of scale, and high transportation costs, uncompetitive in most cases. Table 7.7 shows a comparison of costs for the most popular landfill gas energy recovery technologies.

7.3.5. Landfill Gas Utilization Case Studies

Even with the economic problems noted above, there are and continue to be successes using landfill gas as an opportunity fuel. Industry sources indicate that successful landfill gas energy recovery projects typically have the following characteristics [23]: experienced, professional management, adequate financing which hallows as much

Table 7.7. Cost Comparisons for Typical Landfill Gas Energy Recovery Technologies

Technology/Use	Capital Cost ($/kW)	O&M Cost ($/kW-hr)
IC. Engine/Electricity Generation	900-1,200	0.013 - .020
Gas Turbine/Electricity Generation	1,000 – 1,500	0.01 – 0.015
Steam Turbine/Electricity Generation	1200 [a]	0.002 [a]
Boiler/Direct Heat	1,000 – 1 500	0.005 – 0.018
Organic Rankine/Heat Recovery	1,000 – 1,500	0.005
Fuel Cell/Electricity Generation	3,2 00 [b]	NA

[a] Original estimate in1993 dollars; updated to 2000 dollars by authors
[b] Original estimate in 1995 dollars, using 1995 technology; updated to 2000 dollars by authors
NA – Not Available

Source: [18]

labor, inventory and supplies as seeded, an abundant landfill gas supply, a favorable local marketplace, situation in landfills that remain active for 5-10 years or more, contracts for gas rights, power or gas sales, and facility use that are solid and of adequate duration, and experienced, continuously available personnel for servicing landfill gas extraction system and energy conversion system. With these guidelines, the following are examples of landfill gas-to-energy conversion projects where landfill operators/owners and government worked together successfully [22-23].

7.3.5.1. Detroit Edison – Riverview, MI.

Detroit Edison Company has been involved in the development of landfill gas-to-energy projects since 1986. The Riverview facility has an output of 6.6 MW. Since it began commercial operation in 1988, it has generated more than 225,000 MW-hr of electricity. The project has operated safely and reliably. Riverview municipal officials have recognized the facility's valuable service and its numerous environmental benefits, including capture of some 4 billion cubic feet of methane that would have been released into the environment. The Riverview facility expects to collect landfill gas (LFG) and produce electricity through the year 2027. Detroit Edison has gone on to pursue similar ventures in California, Florida, Texas, Ohio and Michigan [12].

7.3.5.2 Emerald People's Utility District – Short Mountain Landfill.

Emerald People's Utility District (EPUD) worked with Lane Country, Oregon and a private investment company to develop a 3.4 MW plant at the Short Mountain landfill. The plant operates at over 97% capacity and provides a profit to EPUD as well as royalty income to the County. Bonneville Power Administration, in turn, credits EPUD's bill for the power generated by the landfill project [12].

7.3.5.3 Fairfax County, VA – I-95 Sanitary Landfill.

Under a unique arrangement, a developer owns and operates the energy recovery facility, but Fairfax Country retains control of the gas extraction wells. The agreement was structured this way because of the County's concerns about migration and odor control. The County operates the well field for the developer, for a fee, and the developer has rights to a set amount of gas. The I-95 energy recovery facility collects 93,400 m^3 (3.3E6 ft^3) per day of landfill gas and uses 8 internal combustion engines to generate 6 MW of electricity for sale to Virginia Power. By adopting a team approach with the developer, the Country gained a state-of-the-art energy recovery plant at no cost and maintained control over their landfill gas system [12].

7.3.5.4 Mountaingate Facility – Los Angeles, CA.

The Mountaingate Landfill was shut down in 1980. Four of the eight canyons filled during that time now support a championship golf course. The Mountaingate control and recovery plant collects 5 million cubic feet of landfill gas per day. Air Products and Chemical, Inc., operators of the collection facility process the gas on site to remove siloxanes and other impurities using a proprietary process and produce a medium calorific gas (~14.9 kJ/m^3 [500 Btu/ft^3]). The purified gas is then piped to UCLA about 7.2 km (4.5 miles) away. UCLA in turn compresses the gas to approximately 3447 kPa (500 psi) and blends it with natural gas. The blend is used to fuel two 14.5 MW combustion turbine generators that provide power for the UCLA campus. No detrimental effects on UCLA's emissions control equipment have been noted in nearly two years of use [12,21].

7.3.5.5 AT&T Plant – Columbus, OH.

The AT&T plant converted its boilers from natural gas to landfill gas in 1993. The landfill gas is transported from a nearby landfill. AT&T estimates that it saved $120,000 by mid-1995 by using opportunity fuel rather than natural gas [12].

7.4. Coke Oven Gas and Refinery Gas

Coke is an essential input to the steel-making process and is produced by heating coal in coke ovens. To make coke, coal is heated in the absence of oxygen to drive volatile matter from it. Resulting from this pyrolysis process is both the solid product—the coke—and a gaseous material of significant economic potential.

7.4.1. Coke Oven Gas Composition

Coke oven gas, a moderate-calorific gas (<11.9 kJ/m^3 [400 BTU/ft^3]), is produced as a by-product of the coking process [24-25]. A facility producing $9x10^5$ tonnes ($1x10^6$ tons) of coke per year will produce about $1.4x10^6$ m^3 ($50x10^6$ ft^3) per day of coke oven gas [25]. Table 7.8 lists the typical composition of coke oven gas.

Approximately 40% of the coke oven gas is used as a fuel in coke ovens. At most steel plants, the remaining coke oven gas is used to fuel equipment such as boilers and reheat furnaces. The boilers supply steam for electricity generation, turbine-driven equipment such as pumps and fans, and for process heat. Because of the dynamic nature of the steel making process, electricity, steam, and reheat demand vary significantly over time. In most plants, during periods of low electric, steam, and reheat demand, some coke oven gas has to be flared. This results in a loss of energy and a potential cost savings. As a result, some steel companies are finding other uses for their coke oven gas to recover this energy.

7.4.2. Case Study – US Steel – Mon Valley Works

In an effort to save energy and reduce costs, US Steel (USS) developed a system at their Mon Valley Works located just outside Pittsburgh, PA that enabled them to use coke oven gas in their blast furnaces [24]. Although other steel makers in North America had

Table 7.8 Typical Raw Coke Oven Gas Composition

Component	Composition, Actual, %	Composition, Dry Basis, %
Water Vapor	47	--
Hydrogen	29	55
Methane	13	25
Nitrogen	5	10
Carbon Monoxide	3	6
Carbon Dioxide	2	3
Hydrocarbons	1	2

Source: [25]

attempted this, USS was the first to successfully use coke oven gas in blast furnaces. USS targeted their blast furnaces as a potential use for the coke oven gas because the furnaces use a significant amount of natural gas. In the steel-making process, natural gas is injected along with "hot blast" through tuyeres (nozzles) into the blast furnaces. In order to use the coke oven gas to displace some of the natural gas, USS had to make modifications to a number of systems.

USS already had a state-of-the art coke oven gas processing and cleaning facility at their Clairton coke plant. The facility processed the coke oven gas until its content was approximately 50-60% hydrogen. More importantly, the sulfur content of the coke oven gas was significantly reduced during the processing, which allowed it to be used in the blast furnace [24].

USS installed three 900-hp compressors and the associated piping to boost the incoming coke oven gas pressure from 68.9 kPa (10 psig) to 379 kPa (55 psig) for injection into the furnaces. Since not enough coke oven gas would be available to completely satisfy the blast furnace injection requirements, USS purchased instrumentation and equipment so that natural gas could be added to supplement the coke oven gas [24].

Modifications were also made to the blast furnace tuyeres that allowed them to successfully use the coke oven gas. The interior surfaces of the tuyeres were modified to withstand the additional heat and added nozzles to the blowpipes through which the coke oven and hot blast were injected [24].

The results of this project were very positive. USS uses the coke oven gas in their blast furnaces and reduces natural gas consumption, eliminates coke oven gas flaring, and helps to reduce their overall electricity costs. USS has been successfully feeding coke oven gas to the blast furnaces for over six years, and annual savings are estimated at over $6.1 million. The total project costs were about $6 million, the payback for the project was less than one year.

7.4.3. Refinery Off-Gas

Another process industry gas of significance as an opportunity fuel is refinery off-gas. This fuel is commonly used in refinery operations, firing process heaters and related equipment. Within the past 20 years it has become a fuel used in combustion turbines and combined cycle power plants. The typical refinery gas composition is shown in Table 7.9. Note that, like all other waste gases used as opportunity fuel, this gas is methane-based.

These gases have been used in numerous combined cycle projects at refineries including the UNOCAL Refinery in Rodeo, California, the ARCO refinery in Watson, California, the Texaco refinery in Anacortes, Washington, and several projects by Shell Oil [26]. Gas turbines have been developed to fire such gases. Further, numerous burner companies have developed specialized burners to handle the different compositions of these gases.

7.5 Wastewater Treatment Gas

Wastewater treatment gas is a first cousin, if not brother, of landfill gas. The digestion mechanisms producing the gas are similar if not identical, and the gas compositions are also quite similar. Anaerobic digester gas, which consists of approximately 60 percent methane and 37 percent carbon dioxide, is generated in many wastewater treatment plants. Releasing anaerobic digester gas to the atmosphere contributes to greenhouse warming since methane is a potent greenhouse gas. Burning it

Table 7.9. Composition of Waste Refinery Gas

Component	Vol %
H_2	25.7
CO	1.5
CH_4	37.4
N_2	2.9
C_2^s	27.5
C_3^s	2.9
C_4^s	1.7
C_5^s	0.4
Total	100
Lower Heating Value (MJ/m^3)	38.8
Lower Heating Value (Btu/ft^3)	984

Source: [26]

off in flame towers or flares results in emissions of conventional air pollutants such as nitrogen oxides and volatile organic compounds and the loss of a potential energy source. Table 7.10 identifies typical parameters for digesters, including a general description of digester gas. Table 7.11 provides detailed gas composition information for a typical digester gas.

Technologies for using wastewater treatment gas, traditionally, focus upon internal combustion engines although boilers have been used at the very largest of wastewater treatment plants. Utilities such as Inland Empire in Chino, California have experimented with numerous other technologies as well. However, many plants simply flare this gas due to low quantities available or problems associated with using this material.

The Hyperion Plant in Los Angeles was one of the first, and largest, to construct wastewater treatment gas-to-energy facilities. Recently, TVA has pioneered the cofiring of digester gas in a large utility boiler, the Allen Fossil Plant of Memphis, TN. In this project, digester gas will displace 8,000 tons of coal annually. The power will be marketed as green power. The project has been accredited by the Center for Resource Solutions. Numerous others have entered this arena as well.

Table 7.10. Typical Parameters for Wastewater Treatment Digesters

Parameter		Value
Gas Production		
	m^3/tonne	303
	ft^3/ton	11,800
Fuel Values and Characteristics		
	MJ/m^3 (dry)	18.62 – 24.21
	Btu/ft^3 (dry)	500 – 650
	CH_4	55 – 75%
	CO_2	25 – 45%
	Water vapor	Saturated
	H_2S	100 – 2000 ppmv
	Siloxanes	Present

Source: [27]

Table 7.11. Typical Fuel Composition of Wastewater Treatment Gas

Parameter		Value
Gaseous Composition (Vol %, dry basis)		
	Methane (CH_4)	64.1
	Carbon Dioxide (CO_2)	31.4
	Nitrogen (N_2)	4.1
	Oxygen (O_2)	0.4
Molecular weight		25.32
Higher Heating Value		
	MJ/m^3	24.11
	Btu/ft^3	647.7
Ultimate Analysis (wt %, dry basis)		
	Carbon	45.21
	Hydrogen	10.17
	Oxygen	40.17
	Nitrogen	4.45
Higher Heating Value		
	MJ/kg	23.28
	Btu/lb	10002

Source: [28]

7.5.1. Digester Gas Utilization at the Inland Empire Utilities Agency.

Recently, Inland Empire Utilities Agency (IEUA) has become an aggressive leader in the use of this fuel. IEUA is situated in the Chino Basin of California, with a population of 700,000 individuals. This population is expected to grow to 1,2 million by the year 2020. Most significantly, this service area has the largest concentration of dairy cows in the world. Currently the basin produces about 900,000 tonnes/yr ($1x10^6$ tons/yr) of corral-dried manure along with 90,000 tonnes/yr (100,000 tons/yr) of manure washwater [29].

IEUA has installed some 33 energy recovery facilities capitalizing upon digester technology. Using digester gas, they have installed IC engines, microturbines, fuel cells, and other technologies as is shown in Table 7.12.

Table 7.12. Inland Empire Utilities Agency Digester Gas Energy Facilities

Plant	Energy Facilities
RP-1	Two 1400 kW Internal Combustion Engines
	One 625 kW Internal Combustion Engine
	Eight 30 kW Microturbines
RP-2	One 580 kW Internal Combustion Engine
	Two 30 kW Microturbines
RP-4	One 500 kW Internal Combustion Engine
	Six 30 kW Microturbines
Desalter	One 1000 kW Internal Combustion Engine
	One 820 kW Internal Combustion Engine
Demonstration #4	Four 30 kW Microturbines
Demonstration #5	Three 30 kW Microturbines
Future RP-5 and New Headquarters	Two 1400 kW Internal Combustion Engines
	One 250 kW Fuel Cell

Source: [29]

Recently, new fuel cell technologies have been tested by other utilities. With fuel costs on the rise for plants across the country, these projects, and several others underway, hold promise for reaping climate change and economic benefits for many communities [30].

7.5.2. Case Study – Yonkers, NY

A new 200kW Phosphoric Acid anaerobic digester gas -powered fuel cell in Yonkers, New York, is one of the first commercial fuel cells to run on digester gas created at a wastewater treatment plant, producing electricity through a chemical reaction rather than direct combustion and emitting considerably less pollution than a conventional power plant. In addition to the direct reduction of methane emissions, the fuel cell generates about 1.6 million kilowatt-hours of electricity annually, with minimal emissions of air pollutants such as NO_x and SO_x. The heat produced as a byproduct of electrical generation from the fuel cell can be used locally at the sewage treatment plant for heating or cooling to further reduce the facility's energy needs. Compared with releasing the anaerobic digester gas directly to the atmosphere, installing a fuel cell with heat recovery reduces greenhouse gas emissions by the equivalent of 6,800 tonnes of CO_2 and reduces criteria pollutant emissions by 8,600 tonnes.

The Yonkers facility is the one of the first commercial fuel cell project in the world to run on anaerobic digester gas produced during sewage treatment. Solutions such as EPA's patented filtration system, which removes contaminants such as sulfides and halogen compounds from the anaerobic digester gas, have made fuel cells powered by anaerobic digester gas possible.

7.5.3. Case Study – Portland, OR

The Yonkers fuel cell has inspired similar projects elsewhere in the nation, such as the anaerobic digester gas-powered fuel cell at the Columbia Boulevard wastewater treatment plant in Portland, Oregon. The Columbia Boulevard project is the first installation of an anaerobic digester gas powered fuel cell in the western United States. By producing its own electric power from the fuel cell, Portland expects to save more than $60,000 a year in energy costs. The electrical capacity of the fuel cell is estimated to be about 170 kilowatts, which should generate about 1.4 million kilowatt-hours a year. A 170-kilowatt fuel cell would offset about 568 tonnes (626 tons) of CO_2 emissions annually.The success of these innovative fuel cell projects in the U.S. has created interest in anaerobic digester gas fuel cell technology in Europe and Japan.

Clean Water Services in Portland has developed a significant program to optimize the use of digester gas as an opportunity fuel. They have evaluated IC engines, fuel cells, microturbines, and gas turbines. Currently they are implementing a microturbine project at the Durham plant, and system optimization at the Rock Creek plant [27].

7.5.4. Economic Parameters Using Wastewater Treatment Gas

Certain economic parameters have generally been established through the engineering studies of CH_2M Hill [27]. Given the technology status today, and an interest rate of 10 percent, their studies indicate that electricity generated using an IC engine would cost about 6¢/kWh with about ⅔ of that cost being capital, and ⅓ being operations and maintenance (O&M) [27]. Microturbine costs are on the order of 8¢/kWh, with about 6¢/kWh being attributed to capital [27]. The economies of scale apparently do not favor microturbines at this time, although they are improving significantly. Combustion turbine costs are on the order of 2¢/kWh using digester gas, while fuel cell costs are still very high at about 16¢/kWh [27]. No costs are given for boilers, however such systems require more gas than is usually available from wastewater treatment plants.

The use of digester gas, then, requires proximity to the source of fuel, along with selection of an appropriate technology for using that fuel. Finally, use of digester gas requires a commitment to the technology.

7.6. Hazardous Wastes as Opportunity Fuels

Hazardous wastes have long been burned both as a means of disposal and as a source of energy. Utilities such as Duke Power and Northern Indiana Public Service Co. have burned tar-laden dirt from manufactured gas plants in both pulverized coal and cyclone boilers. Many utilities are permitted to burn their own waste oils. Manufacturing industries have burned their own wastes as well. Over the past 30 years, hazardous wastes have entered the opportunity fuels arena, being supplied to cement kilns and industrial boilers. This practice has led to the promulgation of the Boiler and Industrial Furnace (BIF) Regulations by the USEPA. Most significantly, this practice has been used by cement kilns and other industries as a means for reducing fuel costs as a component of product manufacturing costs.

Two types of hazardous wastes typically have been introduced into opportunity fuel commerce: spent solvents and other relatively light organic fluids, and waste oils from automotive and manufacturing sources. Both are widely used in boilers and kilns—particularly cement kilns—throughout the USA and the world.

7.6.1. Fuel Characteristics of Hazardous Wastes

The combustion of hazardous wastes used as opportunity fuel is governed by the BIF rules, and by the Resource Conservation and Recovery Act (RCRA). Dellinger et. al. [4] has characterized representative hazardous wastes with respect to calorific value and pyrolysis kinetics, as shown in Tables 7.13 and 7.14. Note that the kinetics developed by Dellinger relate to the actual temperatures associated with the hazardous molecules. Note, also, the high reactivity associated with these compounds. They readily pyrolyze to fragments and radicals, ready for oxidation.

The Linde Division of Union Carbide Corporation, now Praxair, developed a series of oxidation kinetic parameters for selected hazardous wastes, as is shown in Table 7.15. Note, again, the high reactivity associated with typical hazardous wastes.

7.6.2. Combustion of Hazardous Wastes in Rotary Kilns

Cement kilns are the most common users of hazardous wastes as opportunity fuels. A cement kiln is the world's largest moving manufacturing machine. Typically, they are large cylindrical furnaces 3.6-7.6 m (12-25 ft) in diameter and 137-305 m (450-1,000 ft) in length.

Table 7.13. Calorific Values for Selected Combustible Hazardous Wastes

Compound	Formula	kJ/kg	Btu/lb
Acetonitrile	C_2H_3N	30,850	13,260
Tetrachloroethylene	C_2Cl_4	4,891	2,140
Acrylonitrile	C_3H_3N	33,195	14,260
Methane	CH_4	55,581	23,880
Pyridine	C_5H_5N	32,776	14,090
Dichloromethane	CH_2Cl_2	7,116	3,060
Carbon Tetrachloride	CCl_4	1,005	432
Hexachlorobutadiene	C_4Cl_6	8,874	3,820
Benzene	C_6H_6	41,986	18,050
Monochlorobenzene	C_6H_5Cl	27,628	11,880
1,2-Dichlorobenzene	$C_6H_4Cl_2$	19,130	8,230
1,2,4-Trichlorobenzene	$C_6H_3Cl_3$	14,232	6,120
1,2,3,4-Tetrachlorobenzene	$C_6H_2Cl_4$	10,925	4,700
Hexachlorobenzene	C_6Cl_6	7,493	3,220
Nitrobenzene	$C_6H_5NO_2$	25,158	10,820
Analine	C_6H_7N	36,544	15,710
Hexachloroethane	C_2Cl_6	1,926	828
Chloroform	$CHCl_3$	3,140	1,350
1,1,1-Trichloroethane	$C_2H_3Cl_3$	8,330	3,580

Source: [4]

Table 7.14. Pyrolysis Kinetics Parameters for Selected Combustible Hazardous Wastes

Compound	Empirical Formula	Pre-exponential factor (1/sec)	Activation Energy (kJ/mole)
Pyridine	C_5H_5N	$1.1x10^5$	100.5
Dichloromethane	CH_2Cl_2	$3.0x10^{13}$	267.9
Carbon Tetrachloride	CCl_4	$2.8x10^5$	108.8
Benzene	C_6H_6	$2.8x10^8$	159.1
Monochlorobenzene	C_6H_5Cl	$8.0x10^4$	96.3
1,2-Dichlorobenzene	$C_6H_4Cl_2$	$3.0x10^8$	189.5
1,2,4-Trichlorobenzene	$C_6H_3Cl_3$	$2.2x10^8$	163.3
1,2,3,4-Tetrachlorobenzene	$C_6H_2Cl_4$	$1.9x10^6$	125.6
Hexachlorobenzene	C_6Cl_6	$2.5x10^8$	171.6
Nitrobenzene	$C_6H_5NO_2$	$1.4x10^{15}$	267.9
Hexachloroethane	C_2Cl_6	$1.9x10^7$	121.4
Chloroform	$CHCl_3$	$2.9x10^{12}$	205.1
1,1,1-Trichloroethane	$C_2H_3Cl_3$	$1.5x10^8$	134.0

Source: [4]

Table 7.15. Oxidation Kinetics for Selected Combustible Hazardous Wastes as Determined by the Linde Division of Union Carbide (now Praxair)

Compound	Empirical Formula	Pre-exponential factor (1/sec)	Activation Energy (kJ/mole)
Acrolein	C_3H_4O	$3.30x10^{10}$	150.3
Acrylonitrile	C_3H_3N	$2.13x10^{10}$	213.5
Methane	CH_4	$1.68x10^{11}$	218.1
Methyl Chloride	CH_3Cl	$7.34x10^8$	171.2
Methyl Ethyl Ketone	C_4H_8O	$1.45x10^{14}$	244.5
Benzene	C_6H_6	$7.43x10^{21}$	401.4
Monochlorobenzene	C_6H_5Cl	$1.34x10^{17}$	320.6
Toluene	$C_6H_5CH_3$	$2.28x10^{13}$	236.5
Vinyl Acetate	$C_4H_6O_2$	$2.54x10^9$	150.3
Vinyl Chloride	$CH_2:CHCl$	$3.57x10^{14}$	265.0

Source: [4]

They are set on a slight incline and rotate from 1 to 4 revolutions per minute. Cement kilns can process up to 300 tonnes (330 tons) of raw material such as limestone, clay and sand each hour [31].

Cement kiln energy recovery is an ideal process for managing certain organic hazardous wastes. The burning of wastes or hazardous wastes as supplemental fuel in the cement and other industries is not new. In the US alone, nearly 30 cement kilns use hazardous waste as a supplemental fuel, thus saving the equivalent of 636,000 m^3 ($168x10^6$ gallons) of oil or 900,000 tonnes ($1x10^6$ tons) of coal [32].

Waste fuels are used in the cement industry worldwide, and emissions have been thoroughly investigated in light of government and environmental regulations. Because of the potential toxic effects on human life by some of the pollutants from cement kilns burning fossil fuels combined with hazardous waste, very stringent regulations have been enacted. For example, Table 7.16 lists the maximum achievable control technology (MACT) limits for cement kilns in the USA [33].

Test results have shown that virtually all of the organics, originating from fossil fuel, oil or waste fuels, are destroyed in the process as temperatures exceed 1,340°C (2,450°F) for the material, and nearly 1930°C (3,500°F) in the gas phase in the cement kiln [33-42].

This high temperature must be maintained for several minutes in order to form the clinker minerals that give the cement its hydraulic properties. The other group of compounds associated with both traditional

Table 7.16 MACT Standards for New and Existing Cement Kilns

Pollutant	Standard
Dioxins/Furans	0.2 ng/dscm TEQ*
Particulate Matter	0.030 gr/dscf
Mercury	72 μg/dscm
Semivolatile Metals (Cd, Pb)	670 μg/dscm
Low-volatile Metals (As, Be, Cr., Sb)	63 μg/dscm
$HCl + Cl_2$	120 ppm_v
CO	100 ppm_v
Total Hydrocarbons	Main – 20 ppm_v By-pass – 10 ppm_v

Source: [33]

fuels and waste-derived fuels are the inorganics. Here, metals deserve primary attention, but the metals in waste fuels are not foreign to the process. These same metals originate in coal, coke, fuel oil and naturally occurring raw materials.

The data available clearly show that the higher the kiln operating temperature, the better the destruction of these organics. These high temperatures combined with the long residence times, make the cement kiln an ideal apparatus for cofiring hazardous wastes with fossil fuels.

Although cement kilns are used to destroy wastes by combustion, they are not hazardous waste incinerators. They do have some common characteristics; however, they also have distinct differences. From a combustion viewpoint at least three key differences exist [43]:

1. When wastes are fed at the clinker discharge end of the kiln, the volatilized material experiences a time-temperature history much more severe than in most incinerators (2200-3000°F for 4-12 sec. compared to 1800-2400°F for 2-6 sec., respectively).
2. The waste-to-oxygen ratios in the exit gases are generally lower in cement kilns than in incinerators (% O_2 is between 2-6% in kilns compared to 4-12% in incinerators).

3. The raw meal preheat zones of the process serve as a "low-temperature afterburner" with a high surface-to-volume ratio

It is the effects of these design and operating parameters that minimize the emissions of toxic combustion by-products.

7.6.3. Waste Oil Use as an Opportunity Fuel

Waste oil is a unique hazardous waste, with a long history of utilization. Typical sources of waste oil include automotive oils, machinery cutting oils and cooling oils, and other sources of lubricants. The opportunities to use this material as an opportunity fuel are worldwide. US waste oil production and consumption exceeds some $4.2x10^6$ m^3/yr ($1.1x10^9$ gal/yr), of which 67 percent is burned as fuel and another 4 percent is re-refined [43]. A significant quantity is generated in

Canada annually as well. Blundell [43] reports that 200,000 – 250,000 m^3/yr (53 – 66x10^6 gal/yr) of waste oil is generated in the Ontario province alone; of this 15 percent is burned in cement kilns, 7 percent is burned in small furnaces, and 27 percent is re-refined. In the United Kingdom, 447,000 tonnes of waste oil are generated annually, of which 380,000 tonnes are used –largely as fuel [44]. Significant attention has been given to this waste disposal problem/energy resource opportunity in such other locations as Bulgaria [45], New Zealand [46], Spain [47], and throughout the European Union. States from California [48] to Vermont [49] are paying particular attention to waste oil, its use as an opportunity fuel, and its proper disposal.

The general fuel characteristics of waste oils are shown in Table 7.17. Note the differences between mineral oil and synthetic automotive oil shown in this table. Note, also, the broad range in properties particularly as associated with mineral oil.

Typical trace metal concentrations have also been measured in waste oils, as is shown in Table 7.18. Note that there are significant differences between typical concentrations in the USA and in New Zealand.

Table 7.17. Representative Properties of Waste Oils

Property	Mineral Oil	Synthetic Automobile Oil
Water Content (wt %)	9.65 – 64.1	1.0 – 15.0
Flash point (°C)	160 - 180	140 – 180
Density (g/cm³)	0.89 – 0.95	0.90 – 0.92
Sediments (wt %)	1.38 – 10.5	0.8 – 23.25
Higher heating value (MJ/kg)	18.90 – 39.2	36.0 – 43.8
Higher heating value (Btu/lb)	8,130 – 16,860	15,480 – 18,830
Cl (wt %)	0.07 – 0.25	0.05 – 0.13
S (wt %)	0.69 – 1.10	0.42 – 1.32
N (wt %)	1.60 – 1.95	0.70 – 2.35

Source: [47]

Table 7.18. Typical Trace Metal Concentrations in Waste Oils from USA and New Zealand (values in mg/kg)

Metal	USA	New Zealand
Lead	1100	82
Arsenic	12	8
Cadmium	1	0.8
Chromium	6	2.6
Zinc	800	249

Source: [46]

There are three basic uses of waste oil as an opportunity fuel: in small space heaters and boilers, in larger boilers, and in cement kilns. Of these, cement kilns are the most prominent due to their continued search for low cost alternatives to coal, oil, and traditional energy sources. In New Zealand, for example, two cement kilns dominate the use of all waste oil in that country. Typical emissions from the combustion of waste oil in various applications are shown in Table 7.19. Note that SO_2 is not shown in Table 7.19, due to its dependency on the sulfur content of the incoming fuel.

Table 7.19. Selected Emission Factors for the Combustion of Waste Oil (values in kg/m^3 and lb/10^3 gal of waste oil burned)

Emission	Small boilers	Space heaters	Atomizing burners
Particulates	7.68 (64)	0.34 (2.88)	7.92 (66)
PM-10	6.12 (51)	ND	6.84 (57)
NO_x	2.28 (19)	1.32 (11)	1.92 (16)
CO	0.6 (5)	0.20 (1.7)	0.25 (2.1)
TOC	0.12 (1)	0.12 (1)	0.12 (1)
HCl	7.92 (66)	ND	ND
Lead	6.6 (55)	0.05 (.41)	6.0 (50)
Arsenic	.013 (0.11)	$3x10^{-4}$ ($2.5x10^{-3}$)	$7.2x10^{-3}$($6x10^{-2}$)
Beryllium	ND	ND	$2.2x10^{-4}$($1.8x10^{-3}$)
Cadmium	$1.1x10^{-3}$($9.3x10^{-3}$)	$1.8x10^{-5}$($1.5x10^{-4}$)	$1.4x10^{-3}$($1.2x10^{-2}$)
Chromium	$2.4x10^{-3}$($2x10^{-2}$)	$2.3x10^{-2}$($1.9x10^{-1}$)	$2.2x10^{-2}$($1.8x10^{-1}$)
Nickel	$1.3x10^{-3}$($1.1x10^{-2}$)	$6x10^{-3}$($5x10^{-2}$)	$1.9x10^{-2}$($1.6x10^{-1}$)

Source: [50]

As a reference point, Table 7.20 compares typical emissions from firing waste oil with emissions from firing #2 medium distillate and #4 industrial oil. This comparison highlights some of the opportunities and issues associated with firing waste oil as an opportunity fuel. Note the similarities between the emissions from firing waste oil from gasoline and diesel engines, and the emissions from firing #4 oil. Note, also, the consistently high trace metal emissions associated with firing waste oils relative to typical petroleum products.

Given the typical emissions associated with firing waste oils, it is useful to consider case studies of firing waste oils in cement kilns [51-53]. Two case studies are presented: firing waste oils in cement kilns in Germany, and firing waste oils in cement kilns in Norway. Both of these case studies demonstrate the significant benefits that can be obtained using waste oil as an opportunity fuel. These benefits exist in all industrialized economies, and can be experienced in both utility boilers and industrial applications such as process heat and space heat applications.

Table 7.20. Comparison of Potential Pollutants in Waste Oil and Fuel Oil

Potential Pollutant	Gasoline engine oil	Diesel engine oil	#2 fuel oil	#4 fuel oil
Sulfur (wt %)	0.36	0.25	0.12	0.19
Nitrogen (wt %)	0.04	0.02	<0.01	0.03
Ash (wt %)	0.54	0.46	0.13	0.55
Halogens (ppmv)	<350	<234	<200	<200
Arsenic (ppmv)	---	---	---	---
Barium (ppmv)	2.73	3.39	<1.00	<1.00
Beryllium (ppmv)	<0.02	<0.02	<0.02	<0.02
Cadmium (ppmv)	<1.51	2.34	<0.25	<0.25
Chromium (ppmv)	3.19	3.91	<2.00	<2.00
Lead (ppmv)	47.23	57.00	<10.00	<10.00
Nickel (ppmv)	<1.40	1.85	<1.20	8.34
Zinc (ppmv)	1161	1114	5.00	9.05
PCB's (ppmv)	<5	<5	<5	<5

Source: [49]

7.6.3.1. Waste Oil in Cement Kilns – Germany.

A study utilizing data from cement plants in Germany was done to estimate the emissions of various metals as a function of waste material [52]. There are 76 cement kilns in operation of which 40 are permitted to use alternate fuels such as tires, waste oil, waste wood, etc. A "typical" cement kiln consisting of a raw mill section, a preheater-rotary kiln section and an cement mill section was used to describe cement production in Germany. Using partitioning factors based on information from operating kilns, a mass balance model was developed for this "typical" kiln. Using this information, elemental distributions were calculated for cadmium, lead and zinc when using waste oils at the maximum allowable rate of 30%. The results are shown in Table 7.21.

The results show that nearly all the trace metals exit with the clinker and destruction and removal efficiency (DRE) numbers for the three metals are estimated to be 99.96% for lead, 99.95% for zinc and 99.94% for cadmium. The use of waste oil as an opportunity fuel, then, provides both a low cost energy source and a means for managing toxic emissions from combustion of this material.

Table 7.21. Metal balances for typical cement kilns in Germany
(values in g/t of clinker)

	Metal		
	Cadmium	Lead	Zinc
Inlet Streams			
Waste oil	0.0635	4.77	31.77
Coal	0.0261	10.43	7.39
Raw meal	0.31	23.25	72.85
Outlet streams			
Clean gas	0.000256	0.0152	0.058
Clinker	0.399	38.43	111.94
Destruction and removal efficiency (%)	99.94	99.96	99.95

Source: [44]

7.6.3.2. Waste Oil in Cement Kilns – Norway.

Another study completed in Norway using a dry cement-processing kiln showed similar results with measuring PAH, PCB and PCDD, chlorine and some heavy metals [53]. Wastes cofired with the coal included "typical" wastes, PCB-wastes and waste oils. Testing was completed twice, once in 1983 and again in 1987. The results of this study indicated that the type of fuel incinerated does not influence the emission or organic micropollutants. In both tests (1983 and 1987) neither the emission of organic micropollutants nor that of the particles was higher during the incineration of hazardous waste.

The results indicated also that the emission of metals and chlorides from a cement kiln do not increase during incineration of hazardous waste. The destruction efficiency of PCB in the kiln was at least 99.99997%. These investigations showed that the emission of particles, PAH and other cyclic organic hydrocarbons are more influenced by the operation conditions than by the fuel burned.

7.7. Conclusions Regarding Gaseous and Liquid Opportunity Fuels

The array of gaseous and liquid opportunity fuels is impressive; it reflects the professional ingenuity of the engineering community in its search for cost-effective solutions to energy cost management. These fuels are a significant complement to the solid opportunity fuels discussed in previous chapters of this book.

The gaseous fuels typically are methane-based gases generated by earth's coalification processes, oil refining processes, or the essential mechanisms breaking down organic wastes. These gases have significant potential in the generation of electricity and process heat.

The liquid opportunity fuels are typically considered as hazardous wastes—a legal characterization—and these fuels are typically spent solvents and other light organics, or waste oils. All of these fuels have significant potential to assist energy managers when the conditions are right. Those conditions include proximity to a source of the opportunity fuel, low cost (or negative cost) opportunity fuel, proximity to the grid if electricity is being generated, and an emissions management system (e.g., particulate controls, acid gas controls) capable of handling the emissions resulting from the firing of these fuels.

7.8. References

1. US Geological Survey, 1997, "Coalbed Methane—An Untapped energy Resource and an Environmental Concern," USGS Fact Sheet FS-019-97.
2. Maisel, D.S. 1985. Methane, Ethane, and Propane. In Kirk-Othmer Concise Encyclopedia of Chemical Technology. John Wiley & Sons. New York. p. 619.
3. Stultz, S.C. and J.B. Kitto (eds). 1992. Steam: Its Generation and Use.40th Ed. Babcock & Wilcox. Barberton, OH.
4. Dellinger, B., J.L. Torres, W.A. Rubey, D.L. Hall, and J.L. Graham. 1984. Determination of the Thermal Decomposition Properties of 20 Selected Hazardous Organic Compounds. Industrial Environmental Research Laboratory, Office of Research and Development, U.S. Environmental Protection Agency. Cincinnati, OH. EPA-600/2-84-138.
5. International Energy Agency, 1994, "IEA Global Methane and the Coal Industry," IEA.
6. Mase, S. and H. Hirasawa, 2002, "Coal Mines and the Methane Gas – Development of Integrated Recovery and Utilization System," World Energy Council, 17th Congress, March.
7. "Gist of Kyoto Protocol," 1997, The Japan Times, December 12.
8. DePasquale, M. and B. Pollard. 1998. A Guide to Coal Mine/Greenhouse Projects. USEPA. Washington, D.C. Contracts 68-D4-0088 and 68-W5-0018.
9. "Minfo 48,"1995, Australia.
10. Mutmansky, J.M. 1999. White Paper: Guidebook on Coalbed Methane Drainage for Underground Coal Mines. USEPA. Washington, D.C. Cooperative Agreement CX824467-01-0
11. Tillman, D.A., 1991, *The Combustion of Solid Fuels and Wastes*, Academic Press, Inc., San Diego, CA.
12. Energy Information Administration. 1997. Renewable Energy Annual, 1996. USDOE. Washington, DC.

13. Energy Information Administration. 2001. Renewable Energy Annual 2000. USDOE. Washington, DC.
14. Energy Information Agency. 2001. Form EIA-860B, "Annual Electric Generator Report – Nonutility." USDOE, Washington, DC.
15. Sandelli, G. J., 1992, "Demonstration of Fuel Cells To Recover Energy from Landfill Gas. Phase I Final Report: Conceptual Study," EPA-600-R-92-007, Washington, DC, January.
16. Doorn, M., J. Pacey, and D. Augenstein, 1995, "Landfill Gas Energy Utilization Experience: Discussion of Technical and Non-Technical Issues, Solutions, and Trends," EPA-600-R-95-035, Washington, DC, March.
17. Thorneloe, S.A., 1992, "Landfill Gas Utilization – Options, Benefits, and Barriers," 2nd Conference on Municipal Solid Waste Management, Arlington, VA, June.
18. Williams, T.D., 1992, "Making Landfill Gas an Asset," Solid Waste and Power, July/August.
19. Anderson, C.E., 1993, "Selecting Electrical Generating Equipment for Use with Landfill Gas," Proceedings of the SWANA 16th Annual Landfill Gas Symposium, Louisville KY, March.
20. Siuru, W.D., 1995, "Researchers Test Fuel Cells to Recover LFG," World Wastes, 38, No. 4, April.
21. NST Engineers Inc., 2003, Mountaingate Landfill Gas Project, Company Brochure.
22. National Renewable Energy Laboratory, 1996, "Landfill Gas Recovery Projects Reviewed by NREL," BioCycle, 37, No. 2, February.
23. U.S. Environmental Protection Agency, "Landfill Methane Outreach Program," EPA-430-F-95-068A, Washington, DC, April 1995.
24. Office of Industrial Technologies, Energy Efficiency and Renewable Energy, U.S. Department of Energy Circular – Steel, Best Practices, December 2000.
25. Platts, M., 2002, "the Coke Oven By-Product Plant, Steelworks, American Iron and Steel Institute.
26. Rao, A.D., D.J. Francuz, and E.W. West. 1996. Refinery Gas Waste Heat Energy Conversion Optimization in Gas Turbines. Proc. International Joint Power Generation Conference, Vol 1. ASME. New York. pp. 473-483.
27. Kitto, B. and D. Green. 2003. Electricity Generation Using Digester Gas at Clean Water Services. Proc. 28th International Technical

Conference on Coal Utilization and Fuel Systems. Clearwater, FL. March 10-13.
28. Tillman, D.A. 1997. Combustion Profile of Allen Fossil Plant Boiler #3 Firing Wastewater Treatment Gas as a Supplementary Fuel. Foster Wheeler Environmental Corporation, Sacramento, CA. For Tennessee Valley Authority.
29. Kitto, W.D., N. Clifton, and E.J. Whitman. 2003. Inland Empire Utilities Agency: Meeting Energy and Environmental Challenges in the Chino Basin. Proc. Electric Power 2003. Houston, TX. March 3 – 7.
30. US EPA, 2002, "Inside the Greenhouse," EPA-430-N-02-002, Spring.
31. "Waste-to-Energy/Cement Kiln Overview," 1993, Cadence Environmental Energy, Inc.
32. "Questions and Answers About Cement Kilns, Their Operation and Their Role in Processing Organic Wastes," 1992, Southdown, Inc. Brochure.
33. Lighty, J.S. and J. M. Veranth, 1998, "The Role of Research in Practical Incineration Systems-A Look at the Past and the Future," 27th Symposium International on Combustion, The Combustion Institute, Pittsburgh, PA.
34. Sun, B., A.F. Sarofim, E.G. Eddings, and D.J. Paustenbach. 2001. "Reducing PCDD/PCDF Formation and Emission from a Hazardous Waste Combustion Facility-Technological Identification, Implementation, and Achievement," 21st Int. Symp. On Halogenated Environmental Organic Pollutants and POPS, Kyongju, Korea.
35. Tillman, D.A., W.R. Seeker, D. W. Pershing and D. DiAntonio. 1991. "Developing Incineration Process Designs and Remediation Projects from Treatability Studies," Remediation, Summer.
36. Eddings, E.E., J. S. Lighty and J. A. Kozinski, 1994, "Determination of Metal Behavior during the Incineration of a Contaminated Montmorillonite Clay," Environ. Sci. Technol., 28.
37. McClennen, W.H., J.S. Lighty, G.D. Summit, B. Gallagher, and J. M. Hillary. 1994. "Investigation of Incineration Characteristics of Waste Water Treatment Plant Sludge," Combust. Sci. and Tech., Vol. 101.
38. Holbert, C.H, and J.S. Lighty, 1999, "Trace Metals Behavior during the Thermal Treatment of Paper-mill Sludge," Waste Management, Pergamon Press, Elsevier.

39. Pershing, D.W., J.S. Lighty, G.D. Silcox, M.P. Heap, and W.D. Owens. 1993. "Solid Waste Incineration in Rotary Kilns," Combust. Sci. and Tech., 93.
40. Senior, C, A.F. Sarofim and E. E. Eddings, 2003, "Behavior and Measurement of Mercury in Cement Kilns," IEEE-IAS/PCA 45th Cement Industry Technical Conference, Dallas, TX.
41. Eddings, E.E. and D.W. Pershing, 1996, "Hydrocarbon Emissions Due to Raw Materials in the Manufacture of Portland Cement,"
42. Hansen, E., D.W. Pershing, A.F. Sarofim, M.P. Heap, and W.D. Owens. 1995, "An Evaluation of Dioxin and Furan Emissions From a Cement Kiln Co-Firing Waste," Waste Combustion in Boilers and Industrial Furnaces, Air & Waste Management Association.
43. Blundell, G. 1998. Used Motor Oil Forum Background Paper: Provincial and State Policies on Used Motor Oil Management. Proc. Policy Forum: Used Motor Oil. Recycling Council of Ontario. May 26.
44. Environmental Resources Management. 2002. Waste Oil Recycling. Department of Trade and Industry. United Kingdom. London.
45. Metodiev, M. 2002. Utilization of Waste Oil-Derivitives. Annual of the University of Mining and Geology "St. Ivan Rilski". Vol 44-45, Part II. Mining and Mineral Processing. Sofia, Bulgaria. pp. 147-149.
46. Woodward-Clyde. 2000. Final Report: Assessment of the Effects of Combustion of Waste Oil, and Health Effects Associated with the Use of Waste Oil as a Dust Suppressant. Prepared for the Ministry for the Environment, Wellington, NZ.
47. Regueira, L.N., J. Rodriguez Anon, J. Proupin, and C. Labarta. 2001. Recovering Energy from Used synthetic Automobile Oils through Cogeneration. Energy & Fuels. 15:691-695.
48. Myers, P. 2003. Hazardous Waste Generation Trends in California: 1997 – 2001. Permitting Division, Department of Toxic Substances Control. State of California, Sacramento.
49. Air Pollution Control Division and Hazardous Materials Management Division. 1996. Vermont Used Oil Analysis and Waste Oil Furnace Emissions Study. Vermont Agency of Natural Resources, Waterbury, VT.
50. USEPA. 1996. Waste Oil Combustion. AP-42.
51. Lamb, C.W., et. al., 1993, "Detailed Determination of Organic Emissions from A Preheater Cement Kiln Co-fired with Liquid

Hazardous Wastes," 3rd International Congress on Toxic Combustion By-Products, Cambridge, MA, June.

52. Achternbosch, M. and K.R. Brautigam, 2001, "Co-Incineration of Wastes in Cement Kilns – Mass Balances of Selected Heavy Metals," IT^3 Conference, Philadelphia, PA, May.
53. Benestad, C., 1989, "Incineration of Hazardous Waste in Cement Kilns," Waste Management & Research, 7.

INDEX

C

D

E

T

U

V

W

Z